Patrick Moore's Practical Astronomy Series

Other titles in this series

Astronomy with a Home Computer

Neale Monks

With 87 Figures

Springer

Cover illustration: Front cover computer image courtesy of Apple Computer. Unauthorized use not permitted.

British Library Cataloguing in Publication Data
Monks, Neale
   Astronomy with a home computer.–(Patrick Moore's
   practical astronomy series)
   1. Astronomy–Data processing–Popular works 2. Astronomy–
   Computer programs–Popular works 3. Imaging systems in
   astronomy–Popular works
   I. Title
   522.8′4
   ISBN 1852338059

Library of Congress Cataloging-in-Publication Data
Monks, Neale.
   Astronomy with a home computer / Neale Monks.
       p. cm. – (Patrick Moore's practical astronomy series,
   ISSN 1617-7185)
   Includes bibliographical references and index.
   ISBN 1-85233-805-9 (alk. paper)
   1. Astronomy–Data processing. I. Title. II. Series.
QB51.3.E43M64 2004
522′85–dc22   2004050425

Patrick Moore's Practical Astronomy Series ISSN 1617-7185
ISBN 1-85233-805-9
Springer is part of Springer Science+Business Media
springeronline.com

Typeset by EXPO Holdings, Malaysia
58/3830-543210   Printed on acid-free paper   SPIN 10935750

Preface

Amateur astronomy can be a surprisingly aggressive field. Newcomers to the hobby will often find their computerized telescopes derided as little more than children's toys, equipped with digital crutches for people too lazy to learn the sky the *traditional* way. Apparently, looking at the Moon from your front porch isn't *serious* astronomy, and what you should be doing is driving three hundred miles out into the countryside and working your way through a catalog of magnitude-fifteen galaxies. Otherwise reasonable people will insist that if you look at the faint fuzzy ball *properly*, you can't help but see the chains of stars running through that globular cluster breathtakingly like a squadron of herring gulls following a fishing trawler. Maybe, maybe not, but with all the hoopla over apochromatic refractors and wide-angle eyepieces, the star charts down to thirtieth magnitude and CCD cameras that cost more than small motor cars, some people have forgotten that amateur astronomers look at the night sky not to do science but *simply for the fun of it*. Moreover, one of the best tools for enhancing that fun is probably sitting somewhere in your house right now: the home computer.

However, beyond CCD astrophotography, astronomy books and magazines tend to ignore this particular adjunct to the hobby. This book is my attempt to rectify this, to put as many ideas and tips into one volume as possible, from webcam astrophotography to writing equipment reviews for astronomy web sites. One of my main aims throughout this book has been to keep everything as accessible as possible, the only common denominator being a telescope and a home computer. Even a go-to telescope, while useful, isn't a prerequisite. This isn't a computer manual either, and while there are tips on using computers more efficiently where it relates to amateur astronomy, there isn't anything on how to install programs or write HTML code. Finally, this isn't a book just for users of any one particular kind of computer; in virtually all cases the projects described in this book can be accomplished equally well with Windows, Macintosh and Linux. There's no one best operating system any more than there is a perfect telescope design.

Many people have contributed freely of their time and experience and, without them, writing this book would have been impossible. Particular thanks goes to the software developers who have shared their programs with me and explained something of the philosophy behind their projects. Chief among these are Milton Aupperle (Outcast Software), Elwood Downey (Clear Sky Institute), Jason Harris (KStars), Stephen Hutson (American Dream Partnership), Steve McDonald (Silicon Spaceships), Paul Rodman (Ilanga Software) and Darryl Robertson (Microprojects). Celestron, IBM, Logitech, Meade, Tele Vue and Vixen have been generous with their time and resources, and their help in supplying images in particular is appreciated. Artist and web designer Michele Kraft shared her experience and advice with me while compiling the sections on web publishing and graphic design. My personal thanks must go to David Schultz, editor of the AppleLust web site, who facilitated many of the opportunities I had to review astronomical software and accessories over the years. Thanks also to my friend and colleague at the Natural History Museum in London, Phil Palmer, who helped with the section on photography, and to John Watson at Springer-Verlag for helping shepherd this project through to completion.

Neale Monks
Lincoln, Nebraska
USA

Contents

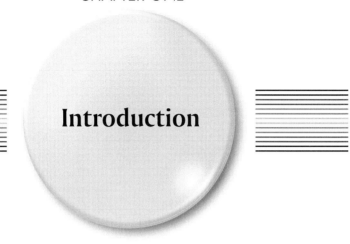

CHAPTER ONE

Introduction

There can be little doubt that computers revolutionized amateur astronomy through the 1990s in the same way as mass-produced Schmidt–Cassegrain (or SCT) telescopes did in the 1970s and high-quality wide-angle eyepieces did in the 1980s. At the time the first reasonably priced SCTs went on sale in the early 1970s, the dominant hobbyist telescope was the equatorially mounted Newtonian reflector. While a fine instrument optically, this design tends to be large, unwieldy and fiddly to maintain. Advocates of the SCT pointed out the much more compact shape of that design coupled with the fork mounting made it a much easier telescope to use, transport and store, and one that required far less maintenance to work well. Critics pointed out that the average commercially produced SCT was more than twice as expensive as a comparably sized Newtonian, and in the opinions of many delivered poorer images. Nonetheless, many hobbyists placed convenience and ease of use over both cost and sheer optical performance, and the 200-mm (8-inch) Schmidt–Cassegrain remains one of the most popular telescopes for intermediate and advanced amateur astronomers the world over. They may not be the best telescopes in terms of sheer optical prowess, but for many hobbyists they occupy a sweet spot as far as balancing cost, convenience and ease of use are concerned.

If the SCT was the big thing in the 1970s, then wide-angle eyepieces such as those produced by Al Nagler's fledgling optics company Tele Vue caught the conspicuous consumption mentality of the 1980s. They were then, and remain now, expensive pieces of kit aimed primarily at advanced hobbyists; for example, the top of the line 31-mm Nagler wide-angle eyepiece costs more than the Meade ETX 90 go-to telescope. However, what these wide-angle eyepieces offered was a new way of looking at the sky. With traditional eyepieces like Plössls, the field of view is *narrow*, that is, around the 50° mark or less, and so relatively low

Figure 1.1. No other telescope design combines performance, price and compactness as well as the Schmidt–Cassegrain telescope (photo courtesy of Celestron).

magnifications are needed if large objects such as open star clusters are to be seen in their entirety. Wide-angle eyepieces have fields of view from 60° to 84°, and so at any given focal length reveal a much larger area of sky than a traditional eyepiece. The result was that it was now possible to view big objects at high magnifications, and so discern subtle details while retaining the visual impressiveness of seeing the entire object at once. Almost without exception, these wide-angle eyepieces worked wonderfully well with the short focal length Newtonians and refractors that became popular at the same time, unlike the more traditional eyepiece designs. Stars were sharp to the edge of the field, and when used on the planets, colors and contrasts were just as striking. In a virtuous circle of supply and demand, many different manufacturers competed to produce ever better eyepieces and telescopes for the increasing numbers of amateur astronomers who were prepared to spend substantial amounts of money to get the best possible views of the night sky.

The SCT had given the hobby a reasonably priced but compact instrument ideal for backyard astronomers, and the burgeoning range of top-notch eyepieces meant that users could get great views of solar system and deep sky objects, but there still remained the problem of finding things to look at. The Moon is an obvious enough target, and so are, generally speaking, the planets, the more interesting of which are usually brighter than any star; but most double stars and deep sky objects are too faint or inconspicuous to be seen with the naked eye, and for beginners, finding these hidden treasures can be difficult. The received wisdom was that the best thing for newcomers to the hobby to do was to learn the sky the old-fashioned way. Essentially this came down to learning the sky as if it were a map. At first, you would begin to learn the names of the brightest stars, and from those you could make out the constellations. Using these patterns to aim your telescope in the right general direction, you would then consult a star chart or atlas to help identify fainter stars to act as landmarks showing the way to the desired deep sky object. This technique, star hopping, works well after some practice and a number of books are available to introduce the technique to those new to it; some favorites of mine are listed in Appendix 1. An alternative method relies on mechanical aids called setting circles. Most commonly, these are discs

Figure 1.2. Although wide-angle eyepieces of various types had been around for years, it wasn't until the 1980s that companies like Tele Vue began to market highly corrected ones providing the flat, sharp field ideally suited to observing star clusters and lunar landscapes (photo courtesy of Tele Vue Optics, Suffern, NJ).

placed around the two axes of an equatorial telescope mount, one marked off in units of declination and the other in right ascension, and so matching the coordinate system used to describe points in the sky. Provided that the telescope is properly aligned to begin with, meaning that the right ascension axis is pointing along the Earth's north–south axis of rotation (i.e., towards Polaris or Sigma Octanis depending on your hemisphere), then the setting circles can be used to "dial up" faint objects without any need to know the sky. In practice though, beginners are apt to find setting circles more trouble than they are worth: on the lower-cost telescopes they tend to be rather crude to begin with, and if the telescope isn't completely stable then the movements in declination and right ascension become too inaccurate to be useful. The more expensive heavy-duty mounts (such as those from Vixen and Losmandy) are much more stable and the setting circles more accurately made and easier to use – but these mounts cost more than the average beginner is likely to spend on a complete telescope set-up, including tripod, mount and eyepieces. Even with the deluxe tripods, between the need for spot-on polar alignment at the start of an observing session and then having to read the small numbers on the setting circles in the dark when moving between targets, many skilled amateurs find setting circles a bit heavy going at times.

Nevertheless, the ability to point a telescope straight at a deep sky target without a detailed knowledge of the night sky is an attractive one, it is just that the skills required to master setting circles take time to learn. Digital setting circles are one way of simplifying the process by offloading the tasks of reading the setting circles and knowing the coordinates of the stars and deep sky objects onto a computer. The computer sits inside a small box and has sensors (called encoders) attached to the two axes of the telescope mount. Once set up correctly, the encoders keep track of where the telescope is pointed and in which direction it needs to move to bring the next object from computer's built-in database into view. The computer does not actually move the telescope, you still need to push the telescope or activate the tracking motors to do this, but it does remove the necessity for the observer to look up the coordinates of objects and try to read the setting circles in the dark. Instead, an LCD display tells the user how much and in which directions the telescope needs to be turned. Having said that, digital setting circle systems never fully lived up to their potential for a number of reasons. For one thing, as with using manual setting circles, if the alignment of the telescope mount is off, or the movement of the two axes is not smooth, then they are pretty well useless. Digital setting circles also tend to be expensive, complicated and unsightly, and the first designs were sold before the computers and encoders were as reliable as they are now, which gave the early digital setting circles a bit of a reputation for capriciousness. A few manufacturers still produce digital setting

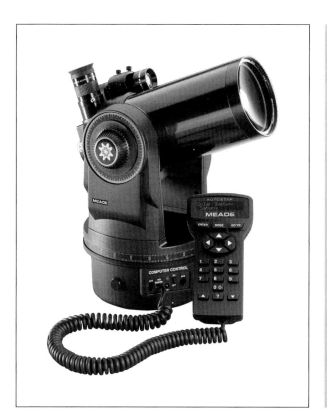

Figure 1.3.
Computerized "all-in-one" go-to telescopes, such as the Meade ETX 90, have quickly established themselves as the standard observing instrument for amateurs of all skill levels (photo courtesy of Meade Instruments Corporation).

circles, such as Tele Vue with their *Sky Tour* package. These are a reliable and mature product ideally suited to alt-azimuth scopes such as the deluxe refractors and giant "light bucket" Dobsonians, but comprise only a tiny proportion of the electronic telescope market. Instead, the dominant telescopes are those that have the computer, mount, encoders and motors all built into a single package. These are the fully computerized, or "go-to" telescopes, that took amateur astronomy by storm in the 1990s.

Released in 1992, the Meade LX 200 series of SCTs were the first mass-produced computer controlled telescopes. Essentially, the LX 200 was a telescope with a computer built into an alt-azimuthal fork mount. The computer needed some input to work, such as where it was located and what the time and date was, and then the user would need to *align* the telescope by pointing the telescope at two or three stars. Through this aligning procedure, the computer would accurately determine the orientation of its altitudinal and azimuthal axes relative to the sky, and this in turn would allow the computer to turn the optical tube of the telescope towards any of the objects in its catalog. The upshot of all this was that provided the user could identify a handful of bright stars to carry out the alignment procedure at the beginning of each observing session, the computer would take care of everything else. The LX 200 delivered on this promise, and it has since been joined in the astronomy market place by a whole horde of computerized go-to telescopes, ranging from inexpensive refractors priced at under a couple of hundred dollars through to massive semi-professional instruments demanding a permanent installation and costing tens of thousands of dollars. Many purists scoff at go-to telescopes because they do nothing to foster the knowledge of the night sky that has been the basis of the science for thousands of years. Others believe they diminish the pleasure of the hobby, which stems more from developing and using the skills involved to find faint stars and galaxies than actually seeing what they look like; after all, much better pictures can be seen in books and on web sites. Whatever the merit to these arguments, go-to telescopes have been amazingly popular with amateurs at all skill levels. This is especially true for people for whom backyard astronomy is a hobby rather than a passion. If you only have an hour or so to observe between dinner and bedtime, then these robotic telescopes are nothing short of a godsend.

Given how popular these go-to telescopes are, and how much they can do, it would probably be worthwhile to write a book about them and expect it to do well, but this book isn't about go-to telescopes, though they are discussed at some length and certainly offer great opportunities for combining astronomy with the digital lifestyle. Rather, this is a book about home computers and astronomy, and that covers a much greater range of activities, projects and skills. Admittedly, a few of the things described in this book work better with certain kinds of computer or telescope designs than others, but provided you have a telescope and some sort of home computer, you'll find plenty of fun and exciting ways to extend your enjoyment of the hobby. The focus of this book is at the low-cost end of the hobby, too, and so while CCDs are one aspect of the hobby home computers have made accessible to amateurs, their high expense puts them outside the topics covered here. Instead, the focus is on projects ignored by the writers of most astronomy books, such as webcam imaging, for which you may well have all the tools needed around the house or loaded on your home computer already.

Finally, although the accent here is on the technology, do not forget that the whole point of backyard astronomy is to enjoy and appreciate the drama and spectacle of the universe. It is all too easy to spend more time looking at a computer screen than through the eyepiece of the telescope. So, sometimes power down the laptop, put away the webcams, wires and widgets, and get back to astronomy done the old fashioned way: just you, a couple of bits of glass, and some photons that left home thousands of years ago.

What Sort of Computer?

Obviously, for this book to be of any value, you are going to need a home computer of some sort. Many people will already have a computer, in which case this section will give a good idea of the sorts of things they can do with whatever machine it is they have. It is central to the philosophy of this book that pretty much any computer running any operating system will expand your enjoyment of amateur astronomy, but it is equally plain that some computer designs and operating systems are better at certain tasks that others.

In determining the sorts of astronomical projects you can do with your computer, a good start is to think about the various accessories that you are going to plug into it. High up on this list should be a webcam of some type; these offer one of the simplest and least expensive ways to get into astrophotography. These cost from $50 to $200 depending on the sophistication of the device and the speed of the interface between the camera and the computer. Most of the lower-priced webcams use the USB 1.1 interface popular on both PC and Macintosh computers although some older webcams use either the PC serial or Macintosh ADB interface instead. Higher performance webcams use either the USB 2 or FireWire bus, but these usually cost about twice as much as a USB 1.1 webcam. Webcams can be used for eyepiece projection photography (where the webcam is held over the eyepiece) but really excel when used for prime focus photography, replacing the eyepiece completely and capturing a high-magnification view ideal for imaging the Moon and planets. An alternative to a webcam is a digital camera, and many people already own one of these long before they think about using them for astrophotography. Digital cameras can work well, though they cost rather more, ranging from a few hundred to well over two thousand dollars, and are somewhat less versatile because they are suitable for eyepiece projection photography only. On the other hand, digital cameras do have some very important advantages. Firstly, with digital cameras there is no need to bring a laptop computer outside as there is when using a webcam for astrophotography. A second advantage of some digital cameras is the ability to alter shutter speed in a similar way to traditional film cameras. The slower the shutter speed, the longer the CCD inside is exposed to light, and the better its images of dim night sky objects are going to be. Unlike most webcams, some digital cameras are therefore suitable for photographing deep sky objects like nebulae as well as the Moon and planets.

If you have a go-to telescope and want to connect it to a laptop computer, you are going to need a serial cable to connect the two together. Go-to telescopes such as those from Meade and Celestron have serial ports built into the handsets or mounts, but these are of the RJ-22 type rather than the more familiar 9-pin serial port seen on most Windows and Linux computers. A converter plugged into the PC will allow the RS-232 serial cable to connect the computer and the telescope. The cable needs to be sufficiently long to prevent tangling while the telescope slews around; a two-meter (six-foot) cable is ideal. Sourcing all these cables is not difficult and none of the kit is expensive, but one reliable supplier that offers all the bits and pieces as well as detailed instructions for a wide variety of specific telescopes models is Software Bisque, the producers of the popular Windows and Mac planetarium program *TheSky*. They make complete packages tailor-made to particular computers and telescopes, including some of the older models like the earlier (pre-Autostar II) LX 200 SCTs and go-to mounts from Losmandy and Vixen among others. One slight complication arises for Macintosh users. Because Macs do not have the 9-pin serial port, a second adapter is required, a USB to serial adapter. These cost around $50, and companies such as Belkin and Keyspan produce Mac OS X compatible models ideal for this purpose.

Software is always a difficult issue when talking about getting the most out a computer, partly because there is so much to choose from but also because software is constantly being updated and improved. Nevertheless, some titles have stood the test of time, and have remained popular and useful for many years. An

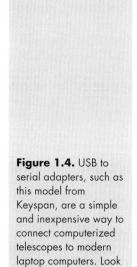

Figure 1.4. USB to serial adapters, such as this model from Keyspan, are a simple and inexpensive way to connect computerized telescopes to modern laptop computers. Look for designs with low power consumption if you want to get the most life from your laptop's battery. (Courtesy of Keyspan. Unauthorized use not permitted).

important factor when selecting software is the operating system of the user's computer. Although there are actually dozens of different operating systems employed by computers all around the world, two dominate the home computer segment of the market. These are the *Windows* family of operating systems (such as Windows 2000 and Windows XP) from Microsoft and Apple's *Macintosh* operating system, known as OS X and is used exclusively by their own line of computers. Of the two, Windows operating systems of one sort or another are by far the most common, and support the widest selection of astronomical accessories and software as well. A third operating system known as *Linux* has also become popular among home users, primarily because unlike the Windows and Macintosh operating systems, Linux costs nothing to download and install. All three of these operating systems can be viable computing platforms for amateur astronomers despite important differences in their design and use (see Table 1).

Desktop Computers

Desktop computers have become the commonest sort of computer used in the home for a number of reasons, but the most important is that they are good value. Because their construction is much simpler than a laptop (which needs to confine all the parts into a very much smaller space), desktop computers are less expensive and easier to maintain and upgrade. Competition between manufacturers is intense and this drives prices down while keeping specifications high. Most desktop computers come with plenty of scope for interfacing with peripherals such as digital cameras, either directly or by adding expansion cards or adapters. Desktop computers normally have nice big monitors, and these are great for working with your astronomical photographs and creating web sites. However, desktop computers do have a serious shortcoming: they are unsuitable for use outdoors alongside the telescope. If your intention is to use a computer to receive images from a webcam or to control a computerized go-to telescope mount, then you need to use a laptop computer instead. While it might be physically possible to drag a desktop computer out into the garden on some sort of trolley and use an extension cable to bring the mains supply out with it, it certainly would not be safe to do so.

Laptop Computers

Laptop (or portable) computers function outdoors safely and easily, but remain relatively expensive compared to desktop computers. This is because they need to be designed much more carefully, not only because they need to be physically smaller, but also to ensure the components use as little power as possible. The batteries and the liquid crystal screens (LCDs) in particular are expensive components to manufacturer to start with – indeed replacing a cracked laptop screen can cost almost as much as buying a new laptop! They are also less expandable and upgrades are much more limited, generally only changing the hard disk and adding more memory are viable options. This is exacerbated by the fact that even

Figure 1.5. Modern desktop computers, such as this IBM IntelliStation M Pro, are equipped with large screens, fast graphics cards and plenty of disk space and memory, making them ideal workstations for processing photographs, creating web pages and running planetarium programs (Courtesy of International Business Machines Corporation. Unauthorized use not permitted).

when brand new, laptops tend to be less powerful than desktop machines of the same vintage. So choosing between a desktop and a laptop comes down to a choice between value for money or convenience. To be fair, though, some of the latest laptops are incredibly powerful and come with large (15 to 17-inch) screens and these really are a viable replacement for a desktop computer.

Many amateur astronomers do choose to buy a laptop simply to run astronomy software and use webcams and CCDs while outdoors with their telescopes, but can opting for a second-hand instead of a new one reduce the cost of a laptop? The good news is that a used laptop can be perfectly useful for running a star charting application such as *TheSky*. Buying second-hand computers is always a bit of a lottery, but there are few tips worth bearing in mind. Most important, bear in mind that retailers often give some sort of warranty (perhaps only a few weeks or months, but that should be enough to time to tell a peach from a lemon!) but if you buy a laptop from a private individual it is very much a case of *caveat emptor*. Secondly, be aware that the batteries that come with second-hand laptop computers retain less charge than when new, so buying a new battery for an old computer is often essential. The batteries in modern laptops are said by their manufacturers to give as much as five hours of use, but even the best battery has a finite life, and after a couple of years will give much less. This will become most noticeable if the computer is working continuously and cannot slow down the internal hard disk, as will be the case when using a laptop to record images from a webcam. An important thing to remember when using any laptop outdoors is that batteries work badly in very cold conditions. Because batteries produce electricity by a chemical process, the colder it is, the slower the chemical processes, and the less the charge produced. A final consideration when choosing a second-hand laptop computer is whether it will work well with your desktop computer, if you have one. Overall, this is much less of an issue than it used to be: if your desktop computer is a Mac, it is perfectly possible to use a PC laptop for your use outdoors, or vice versa. Swapping data between the two is easy, thanks to shared standards for things like network protocols, Zip disks and CD-ROM formatting; and many pieces of hardware including webcams and digital cameras, are fully compatible with both operating systems.

What About Handheld Computers?

For many people the ideal computer for a backyard astronomer is a laptop, but a possible low-cost alternative is a handheld (or *palmtop* computer). These come in many different shapes and sizes, running from basic ones that are little more than glorified personal organizers through to machines with color screens, built-in networking and significant amounts of processing power. Two operating systems predominate, the *Palm OS* designed specifically for low power, small screen computers, and a lightweight version of the Windows operating system called *Windows CE*. Both are popular and commonly seen, although the Windows CE operating systems is more often seen on the more expensive and powerful handheld computers (if only because it demands significant amounts of power and screen space to run). Some of the top-end models can run an impressive array of astronomical software ranging from relatively simple applications for determining the phase of the Moon through to full-blown planetarium programs similar to those that run on desktop and laptop computers. One of the most impressive astronomical software packages is from Software Bisque, the producers of the popular Windows and Macintosh planetarium program *TheSky*. They produce a Windows CE version called *TheSky Pocket Edition* for handheld computers. It features star charts, phase diagrams of the planets, positions of the moons of Saturn and Jupiter, and many other useful functions. By connecting the handheld computer to a compatible go-to mount *TheSky Pocket Edition* can even be used to control the telescope. A serial cable connects the go-to mount with the handheld computer, plugged directly into the computer or into a docking cradle into which the computer rests, depending on the design. Palm users can enjoy applications with a similar feature set such as Andreas Hofer's *Palm Planetarium*, Kevin Polk's *2Sky*, and others. Clear Sky Institute, producers of the desktop computer program *XEphem*, also produce a star-charting program for Sharp Zaurus palmtop com-

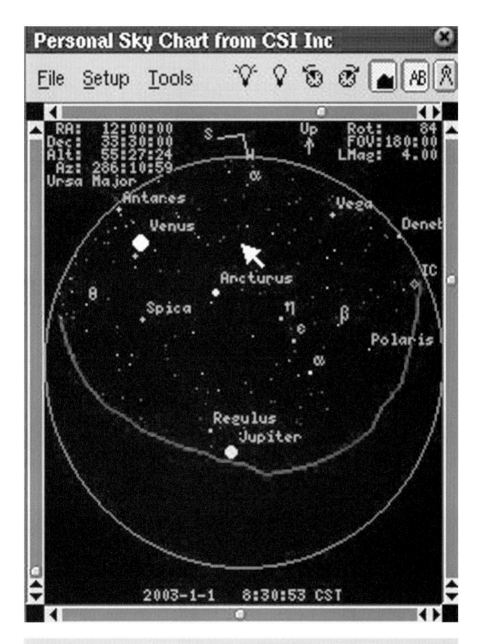

Figure 1.6. Planetarium programs exist for a variety of palmtop computers, such as *Personal Sky Chart* for the Zaurus platform. If your needs are restricted to identifying bright stars or locating showpiece deep sky objects, these devices provide an inexpensive and highly portable alterative to a full-sized laptop (screenshot courtesy of Clear Sky Institute).

puters called *Personal Sky Chart*. Web archives are the main source of astronomical software for handheld computers, with little or nothing sold in retail stores. Most of the software is shareware, meaning that while users are free to download and try out the software without obligation, if they decided to keep the software they have to pay the registration fee. Generally, users download software using a desktop computer and then upload it from the desktop computer to their handheld computers.

What these handheld computers generally cannot do is support astrophotography equipment like webcams, although the ones with color screens will display images. In fact, a slide show of deep sky photographs on one of these little machines would be a great way to supplement a sidewalk astronomy session with the neighborhood, giving people the chance to see something like the Hercules Cluster with their eyes through the telescope first, and then let them see how Hubble sees it on the handheld! The fact that these handheld computers are so small and portable, and last so long on one set of batteries, puts them streets ahead of laptops as far as convenience is concerned.

Which Operating System?

Three operating systems are in wide use among amateur astronomers, Windows, Macintosh and Linux, and all three are capable of doing all of the projects described in this book. Only two are as easy-to-use, consumer level products, Windows and Macintosh, and of these, the various flavors of Windows are by far the most popular among home users. Unsurprisingly then, the availability and diversity of software and accessories is greatest for computers running Windows, and without exception devices like CCDs and go-to telescope mounts are either Windows compatible straight out of the box or come with software to make them so. Consequently, the default computer of choice for the average amateur astronomer is a laptop running a modern version of Windows. Having said that though, there is still a huge range of software for computers running the other two operating systems, and indeed each operating system has distinct advantages over the others that could be important factors. So, which is best? This question causes some of the most passionate arguments among geeks and technophiles, arguments that generally produce more heat than light.

Microsoft Windows

Computers that run one of the different versions of the popular Windows operating system (such as Windows XP) are referred to as *PCs*. This is an abbreviation of *IBM-compatible Personal Computer* and a reference to the fact that although a variety of manufacturers build PCs, they all adhere to an overall set of standards and compatibilities and so can use the same peripherals and software. These are by far the most common personal computer found in the home, and consequently among the most popular for manufacturers of astronomical hardware, software and utilities.

Table 1.1. Windows, Macintosh and Linux platforms compared

	Range of commercial astronomical software	Range of shareware/ freeware astronomical software	Compatibility with go-to telescopes	Range of commercially available webcams	Range of commercially available digital cameras	Range of commercially available CCDs
Windows	Very wide, including *TheSky*, *Megastar*, and many others	Very wide, including planetarium programs (such as *Cartes du Ciel*), Moon-mapping software, utility, and astrophotography utilities	All go-to telescopes and mounts are Windows compatible	All webcams include Windows compatible software	All digital cameras come with Windows compatible software	Most amateur level CCDs are designed solely for use with Windows computers
Macintosh	Fewer than for Windows but still good, including *TheSky* and *Starry Night*	Smaller range than for Windows, though good planetarium programs, telescope control, and astrophotography software exists	Although not designed to be Mac compatible, most mounts and go-to telescopes are, provided a USB to serial adapter is used to make the connection	Many webcams are Mac compatible out of the box, and some others can be made so using third-party software	Many digital cameras come with Mac compatible software	Only a few CCDs are Mac compatible, although many others will work with Windows emulation software such as *VirtualPC*
Linux	Limited, e.g., *XEphem*	Very wide, including *KStars* planetarium program and various utilities and graphics programs	Possible but less easy than with either Windows or Macintosh, e.g. *XEphem* can be used to control LX 200 telescopes	Few if any webcams are sold as Linux-friendly though they can be made to work using third-party software	No digital camera comes with Linux software, though with third-party software pictures can usually be uploaded from the cameras to the Linux computer	Very few CCDs are Linux compatible, though third party software exists for a few

Pros:

- *Intense competition between the many producers of PCs has kept prices low and specifications high.* Many manufacturers offer incentives to attract purchasers, ranging from bundles of free software through to useful accessories such as digital cameras, webcams, scanners and printers. A careful shopper can get a good price on a powerful, reliable package.

- *The widest range of software is available including some of the most popular and well-respected programs.* These include Patrick Chevalley's powerful but free *Cartes du Ciel* program; popular commercial packages such as SPACE.com's *Starry Night* and Software Bisque's *TheSky*; and heavyweights like Willman–Bell's *MegaStar* and Chris Marriott's *SkyMap Pro*, one or other of which is used by many advanced hobbyists. Many of the multi-media compact discs that come with astronomy books are Windows compatible, and only sometimes usable by other operating systems.

- *PCs benefit from a huge variety of Internet resources.* Besides numerous pieces of shareware and software, the majority of the web sites that offer advice on webcam imaging and interfacing computers with astronomical hardware are oriented exclusively to users of the Microsoft Windows operating system.

- *PCs enjoy off-the-shelf compatibility with virtually all pieces of astronomical hardware.* Go-to telescopes and mounts are designed with Windows PCs in mind and the two can be connected without the needed for additional software or adapters. Even if you do not use a laptop to control the telescope, this is useful for updating the software used by the Meade Autostar handsets, where updates to the software in the handset can be downloaded using a PC.

- *PCs are fully compatible with all webcams, digital cameras and astronomical CCD cameras.* Webcams and digital camera manufacturers naturally enough cater for the huge market of home users with Windows PCs, so finding a compatible webcam or digital camera is easy. Moreover, the producers of astronomical CCDs design and build their cameras almost exclusively for Windows PCs. Together with third-party developers, they also offer an impressive range of image editing and utility software, such as Meade's *Epoch 2000ip* CCD image processing software, Robert Stekelenburg's *AstroStack* for optimizing webcam and digital camera images, and Starlight Xpress' *STAR* program for autoguiding.

Cons:

- *PCs can be difficult to use.* The sheer diversity of PC producers using components supplied by thousands of other companies makes it impossible for the default installation of Windows to guarantee support for any given component or program in a computer. It is quite common for new pieces of hardware to need extra software (called "drivers") to work, or for programs to interfere with one another and solving these conflicts is extremely tedious.

- *You get what you pay for.* A bargain PC could be exactly that, a bargain, but it could equally easily be an obsolete or difficult to upgrade computer bundled with box-loads of flashy but poor quality software and accessories. If you aren't familiar with computers, it is a very good idea to bring a friendly computer geek shopping with you to help sort the wheat from the chaff!

- *After-sales service varies enormously.* Computers are complex machines and they do go wrong, so good after-sales service is essential. Choose your source carefully: sometimes it is better to pay a little more for the goods if that means buying from a retailer with a good return and refund policy. The best retailers will take back faulty equipment without fuss and either replaces it at once with new stock or promises to repair the equipment within a few weeks. On the other hand, though offering the lowest prices, mail-order suppliers can be awkward when it comes to getting faulty machines repaired or replaced, so establish beforehand how they deal with returns and refunds, and if the warranty that comes with the goods demands that your return the computer to the manufacturer instead of the retailer.

Apple Macintosh

Unlike the situation with Windows PCs, which combine hardware from one or more vendors with an operating system from Microsoft, Apple Computer builds both the hardware and the operating system that makes up a Macintosh. Overall, Apple Macintosh computers have a reputation for being well designed and reliable, if slightly more expensive than a comparably appointed Windows PC, but there is no escaping the fact that Apple is a niche player in the home computer market. Nonetheless, many of the major astronomy software developers produce Mac as well as Windows versions of their programs, and many of the accessories popular with amateur astronomers such as webcams and go-to telescopes are Mac-compatible. The Mac operating system, OS X, is UNIX based, and a number of Linux astronomy applications will work on the Mac with only a little modification. One slight complication to buying Mac software and accessories is the existence of "legacy" software that runs only in the older Mac operating system, referred to now as the Classic OS, but in its time as the MacOS. Running them is not the problem on modern Macintosh computers (OS X runs Classic applications like *TheSky* and *Voyager III* just fine); the problem comes when connecting these older applications to things like webcams or go-to telescopes. While an OS X program can communicate with external hardware perfectly well, Classic programs running on an OS X computer often fail to.

Pros:

- *Macs are easy to use.* The Macintosh operating systems is widely considered the easiest operating system to use. Installing new software, expansion cards and other accessories, and upgrading things like hard disks, is usually straightforward and easy to do. This is because Apple Computer makes both the computer and the operating system, so it can more fully test and troubleshoot software and hardware, and keep a tight control on the standards and protocols used by third party developers. This contrasts with the situation with Windows and Linux, where a variety of manufacturers produce the computers and it is very difficult for any one developer to test their products on every single possible hardware configuration.

- *Wide range of astronomical software.* Although smaller than the range of titles available for Windows, Mac users still have access to a very wide range of different programs. Indeed, the two operating systems have many programs in common, including *Starry Night, TheSky, XEphem* and iLanga's *AstroPlanner*. There are Mac-only applications as well, though rather fewer than for Windows. Among these are planetarium programs like Syzygy's *The Digital Universe*, Southern Star's *Sky Chart 3*, Carina's *Voyager 3* and Microproject's *Equinox*; utilities like Stephen Hutson's *ScopeDriver* for planning and running observing sessions using go-to telescopes; and Keith Wiley's *Image Stacker* that stacks and processes webcam movie frames to make high-quality images.

- *Mac OS runs UNIX applications natively.* The Macintosh operating system is UNIX based, just like Linux (see below), and in theory at least this gives Mac users access to a whole range of astronomical applications including those used by professional astronomers. Linux titles ported to OS X include *XEphem*, the KStars Team's *KStars* and Fabien Chéreau's *Stellarium*. Among the professional-grade UNIX applications that amateurs could find useful (or at least interesting) are *fv*, a FITS file viewer from the NASA High Energy Astrophysics Science Archive Research Center; the US National Optical Astronomy Observatories' *IRAF* image analysis application; and the Smithsonian Astrophysical Observatory's *DS9* data visualisation program.

- *Macs are compatible with much astronomical hardware.* Many USB and FireWire webcams come with Mac-compatible software, and the popular LX 200, Autostar and NexStar go-to telescopes are fully Mac-compatible, although serial port adapters may be required. Modified webcams optimized for astrophotography like Sonfest Promotion's SAC-IV cameras work with Mac computers (although third-party software may be needed, such as webcam drivers from Logitech and 3Com or image capturing software like *ReelEyes* from iREZ).

Cons:

- *Macs enjoy limited CCD camera support.* This is the single biggest weakness that Macs have as far as amateur astronomy is concerned. Most astronomical CCD producers, including Apogee, Celestron, LISÄÄ and Meade, do not currently offer any Mac OS software or support. There is *some* hope though for Mac users wanting to get into CCD imaging, though. British CCD manufacturer Starlight Xpress is supporting third-party development of drivers for its USB CCD cameras; and the Santa Barbara Imaging Group (SBIG) are promoting third-party development of drivers for their cameras, particularly the USB ones, and have gone so far as to release a beta version of their *CCDOps* software. SBIG have historically been one of the few companies to offer Mac OS software for their cameras, and some of the older ones, including the ST-4X, ST-5 and ST-6 are Mac-compatible. SBIG also confirm that Microsoft's *Virtual PC* program emulates the Windows operating system sufficiently closely for their USB ST-7X family of CCDs to be used that way as well. Mac users who want to use non-USB CCD cameras can use a free Mac program called *SkySight* from Southern Stars (producers of *Sky Chart 3*; the link to the web site is in Appendix 1 along with other planetarium programs) that allows Macintosh computers to use a number of otherwise PC-only CCD cameras. The following cameras are supported: the Celestron PixCell 237 and 255; the Meade Pictor

208 and 216; the SBIG ST-5, ST-5C, ST-237, ST-6, ST-7, ST-8 and ST-9; and the Starlight Xpress HX-516, MX-516, MX-916, MX-5 and MX-5C. Jeff Terry is developing a program called iCCD that allows Mac users to run modern CCD cameras including the MX-7, MX-7C and SXV-9C. This program is a free download from his web site: http://mrmac.mr.aps.anl.gov/~astronut.

- *Not all programs come in Mac as well as Windows versions.* While Windows and Mac users both enjoy their own versions of many programs, there are some important and popular programs that are Windows only. At the high end, *MegaStar* and *SkyMap Pro* are two notable gaps in the Macintosh software roster, and less ambitious observers on a budget will miss the free but powerful *Cartes du Ciel.* One solution is to use Windows emulation software to run these applications if you absolutely must, but for the most part there are Mac alternatives to Windows-only programs (this is discussed in Chapter 3).

Linux

Linux is another UNIX based operating system like Mac OS X; but unlike OS X, it is not a privately owned and trademarked product, instead being an "open source" project that a worldwide community of volunteers write and improve. Linux developers then share their improvements, incrementally developing the software and responding to bugs and problems as they arise. In many ways, it is the computer geek's dream: there are plenty of things to fiddle with and tweak, and although graphic user interfaces exist (such as *GNOME* and *KDE*), many people enjoy getting into the real nuts-and-bolts of the operating system via a command-line interface known as a terminal. Because it is a UNIX operating system it is lean, fast and stable; but the real attraction for many is that Linux runs well on even relatively old hardware and if you download it from the Internet, costs nothing to install. Linux can be great way to breathe new life into an old computer at minimal cost.

Pros:

- *Software, including the operating system, is low, even zero, cost.* All the software, from the basic operating system through to the astronomical specialties like planetarium programs and webcam drivers, is available on the Internet. Of course, this can take hours even with a fast cable connection, and probably days with a modem. You would also need a working computer (be it a Mac or Windows PC) to download these files in the first place. Nevertheless, the option is there, although it is usually easier in most instances to by pre-packaged CD-ROM installations of the Linux operating system and utility software. Various companies produce tailor-made packages for different hardware configurations, such as Red Hat Linux for PCs and Yellow Dog Linux for Apple hardware. Even bought off the shelf, Linux usually works out much cheaper that buying Windows or Mac operating system packages, generally a few tens of dollars rather than hundreds as is the case with the two mainstream operating systems, and you get a helpful manual and other resources in the box as well.
- *Linux runs a very wide range of scientific-grade UNIX applications natively.* Professional astronomers generally use UNIX computer workstations and

many of their favorite programs install and run on LINIX machines with little or no effort. These include *fv* and *IRAF* mentioned earlier as well as many curious and interesting pieces of software such as those capable of driving professional (as opposed to consumer-grade) CCD cameras; mathematical models for simulating variable and binary stars; tools for carrying out radio and x-ray studies, and more. Not all of these would be terribly useful to the average backyard astronomer but some are certainly interesting and fun to explore.

- *Linux users enjoy a viable range of consumer-level astronomical software.* Compared with either Windows PCs or the Mac, the range of commercially produced and distributed Linux astronomical software is limited. The heavyweight astronomical ephemeris application *XEphem* is an exception, and available in Windows, Mac and Linux versions. To some degree the paucity of commercial software is offset by a good number of useful open-source software projects including star charting and planetarium programs (such as the relatively lightweight *KStars* planetarium), drivers for using webcams and CCDs; and image processing programs like the *GIMP*. Random Factory (www.randomfactory.com) sell a "Linux for Astronomy" CD-ROM that includes a selection of these programs that is a convenient starting point, but practically all Linux astronomical software can be downloaded from the Internet and used for free.

Cons:

- *Linux is not easy to use.* While the usability of Linux for ordinary home users continues to improve, there are still aspects of the system that require knowledge of the fundamentals of UNIX and the command line. In particular, installing new software, setting up printers and so on can be tricky.

- *Linux versions of popular Windows programs are lacking.* Although there is some astronomical software available, the mainstream titles like *Starry Night* and *TheSky* are not available in Linux versions. Even downloading and installing Linux software is more complicated than for Macs and Windows machine because software comes as source code that needs to be "configured" and "compiled" specifically for your hardware and software combination. This is not a trivial operation, and if you do not have experience and knowledge of this sort of thing, then you will find the whole process completely baffling.

- *Much open-source Linux software is perennially beta.* Often, open source software (including much of the software available for Linux) is as an ongoing project rather than a fully polished product. Consequently, Linux software can be buggy or unstable on certain hardware configurations, and usually lacks the user-friendliness of commercial Windows and Mac software.

- *Limited CCD support.* As with the Macintosh, support for astronomical CCD cameras is limited and none of the CCD camera manufacturers offers Linux software for their devices. There are third-party drivers and utility software, however, for a number of popular CCDs. Random Factory produce commercial Linux software for the Starlight Xpress MX5 and Apogee CCDs.

- *There is very little commercially produced hardware.* While it is easy to go into a computer store and buy a Windows PC or an Apple Macintosh computer, buying a Linux machine is more complicated. There is no single mass producer of Linux machines in the way that Apple Computer is for Macintosh, and while a few companies produce versions of their PCs with Linux installed (such as IBM) these are generally aimed not at home users but for the business end of the market for use as web servers. Some retailers modify off-the-shelf PCs by replacing the Windows operating system with Linux, but more commonly, Linux users buy Windows or Mac compatible hardware and install Linux onto it by themselves. This is a time-consuming process though not intrinsically difficult for those who are comfortable tweaking or upgrading computers generally.

CHAPTER TWO

Modern Amateur Astronomy and the Internet

The sheer scope and abundance of the information available on the Internet has proved to be a great boon to the amateur astronomer. In this chapter, we will be looking at web pages, mailing lists and newsgroups in turn and examine the best way to get useful information out of them, and most effectively contribute your own experiences and successes so that others can benefit from them.

Web Pages

Publishing on the Internet is so quick, simple and inexpensive that any individual or astronomical society can put up notes, observations and opinions for all to see. Many individuals construct web sites around their hobby, some giving accounts of their observing tools and goals, while others choose a more educational approach explaining the basic methods or suggesting activities. A number of societies exist that concentrate on observing a single type of astronomical object, such as variable stars or the Moon, and their web sites aim both to encourage participation by others in that field of study and to publicize the work down by the group so far. The best of these sites produce calendars of astronomical events and maps for finding interesting deep sky objects, helping to bring new people into the ranks of astronomy hobbyists. Some societies and clubs even offer certificates and other incentives to amateurs completing a given list of observations.

As well as contributing to groups, many amateurs like to display their astronomical images on their personal web pages, and indeed this is one of the best ways to share the fun of your hobby with your friends. Though not many of the people you know would really want to stay up until one in the morning on a freezing cold night just to see Jupiter, they may well find a few of your pictures a fun diversion while drinking their morning coffee at work. We'll be looking at the best way to publish your results on the Internet later in this book when we discuss digital astrophotography, but for now our attention turns to finding, appraising and writing astronomical equipment reviews.

Equipment Review Sites

Equipment review web sites of one sort or another have been incredibly popular, and many amateurs turn to the Internet for information on making new purchases long before they being to talk with their retailers. Astronomical equipment is expensive (a good, basic reflector will cost at least a two or three hundred dollars, while top-quality large aperture apochromatic refractors cost as much as a small motor car) so unsurprisingly newcomers to the hobby are eager to get the best value for their money. Besides being expensive, astronomical equipment is also complicated, with a bewildering array of designs, not just for the telescopes but also for things like eyepieces, binoculars, mounts and software. Appendix 1 includes some books that discuss astronomical equipment in depth, fairly and objectively. However, books are updated every couple of years, at best, and even then, they cannot possibly cover every piece of kit available. So where can prospective purchasers turn to ask detailed questions about pieces of equipment?

They can of course ask their questions to the staff at their local telescope retailer. A dealer that specializes in astronomy rather than cameras or scientific toys should be able to offer good advice, particularly if the sales clerks are experienced astronomers themselves. However, a problem arises if you want advice on the relative merits of certain design of telescope, say a 200-mm (8 inch) SCT, made by two or more competing manufacturers, a situation not beyond the realms of possibility. A large number of dealers have undertakings with manufacturers to sell just one brand of telescope (such as Meade but not Celestron) and their knowledge is primarily about that particular brand. It is unlikely that such dealers would recommend equipment sold by their rivals on the other side of town. A different sort of problem exists with popular astronomy magazines, which try to maintain good relations with their advertisers. Many hobbyists suspect that the reviews published in these magazines are somewhat moderated by the importance of maintaining this important source of revenue. Slam a manufacturer's new telescope, eyepiece, or some other piece of kit too hard, and they might choose to advertise elsewhere. Therefore, although retailers and magazines can be useful, experienced amateurs recommend that newcomers seek advice from other astronomers, such as at a local astronomy club. While this remains a very useful approach, for many people who do not belong to an astronomical club, perhaps because there is not one in their town, the Internet has become an alternative channel for owners and users of equipment to share their experiences with others.

The lack of commercial or editorial biases and the vast number of contributors should make the Internet the most comprehensive and objective source of astronomical equipment reviews. Equipment review sites basically fall into two types: single author web sites where one person carries out all the reviews, and multiple author web sites where reviews from many different people are edited and posted. Appendix 1 includes links to some good examples of both type, and any of these are a great place to get information about equipment before laying down the cash at your retailer. Each type does have its advantages and disadvantages, though. Single author web sites are obviously the work of one person, and as such are tend to be a labour of love, growing slowly but surely. Because one person is doing all the reviews, the individual reviews are broadly comparable in approach and the criteria used to judge value, quality and performance. This makes it much easier to identify the best products of each class, and after all, many people turn to reviews so that they can make decisions about which of a bunch of competing models of telescope or whatever is the best to get. On the other hand being a smaller, more slowly developing project, limited by the time and resources available to the single reviewer, these sites take a long time to become comprehensive, and many never really do. Personal web pages for example often have a just a handful of equipment reviews, a couple of telescopes and a few eyepieces, nestling with the rest of the astronomical content, and the value of such sites as a source of information on products is naturally much more limited. If a reviewer has only ever used a single Newtonian how will they know how it compares to similar aperture Newtonians from other manufacturers?

Figure 2.1. *Cloudy Nights* is one of the best review web sites for hobbyists of all levels of experience and disposable income (screenshot courtesy of Cloudy Night Telescope Reviews).

In contrast, multiple author web sites grow very fast because they are not limited to a single person's resources or time. Instead, many people all review their own equipment and send the results in, and an editor pulls them altogether to build the full web site. This division of labour obviously speeds the operation up enormously, but it does mean that reviews are much more heterogeneous in quality and structure. Some authors will be experienced, other much less so, and criteria such as optical performance will be judged in different ways by different people. This all means that drawing conclusions on a particular range of products from a selection of reviews by different people can be tricky, even unsafe. If you are looking to choose a 32-mm Plössl, and there are three brands reviewed on the web site, each by different people, you want to be sure that each review is actually comparable. A solution to this adopted by the best multiple author sites is *peer review*. Instead of immediate editing and posting, a team of experienced amateurs read the review first. Once they've looked at the review, they can offer criticisms and ideas for improvement and standardization, allowing the author to revise and enhance the review before its final editing and publishing. By doing this with all the reviews, a multiple author web site could aim towards getting the consistency of quality that characterizes a single author site, but at the fast pace of development typical of multiple author sites, essentially the best of both worlds.

Objectivity and experience are the two fundamentals to producing good reviews. Some authors come to their reviews with years of experience and having tried out many different brands and designs of telescope, eyepiece, computer program or whatever. Such authors make balanced comparisons with other products in the same market and price range. An example of a useful and objective statement might be that "the field was flatter at the edge of the field of view and there was definitely less ghosting in this Brand-A 32-mm Plössl than in the Brand-B 32 Plössl that came bundled with my telescope." Note that it does not take any level of expertise to declare that a two thousand dollar apochromatic refractor has better color correction or resolution than an achromatic refractor of similar size that cost a tenth as much. Neither is it a surprise that a 12-inch Dobsonian gave a bigger, brighter image of a globular cluster than an 80-mm wide-angle refractor. These sorts of observations are a given: Some designs perform better than others do, price is generally directly proportional to quality, and advantages in aperture overrule practically everything else. Real reviewing is about discriminating between products of the same type, design or function, in such a way that the best examples can be marked out for potential purchasers as good value; and to do this well takes experience. However, many reviews come from amateurs writing about their first telescope, or with only a limited sampling of other designs and models of products of a certain type. A statement like "this is the best telescope I have looked through" is obviously not terribly useful if based on a couple of years of experience with only two or three different telescopes.

Equally misleading are reviews from writers who clearly have an axe to grind as far as the products of one particular manufacturer are concerned. Statements like "no product from manufacturer X should ever be bought" are only useful if based on a fair sample of that manufacturers merchandise. Certainly there are manufacturers that produce what are basically toy telescopes and useless for serious hobbyists, but equally the sometimes vehement advocacies of one brand over another that can be found in many reviews and articles isn't necessarily a balanced point of view. However, by far the most difficult and insidious bias to spot is where the writer holds one particular design of telescope or accessory above all others to the point of almost religious devotion. To read some articles on the Internet, you would imagine that a 102-mm (4 inch) apochromatic refractor would walk all over a 200-mm (8 inch) SCT as far as optical performance is concerned; of course it can't. A well-made refractor may deliver nice, sharp images, but it can't magically produce brighter, more resolved images at higher magnifications than a reflector with twice the aperture and four times the light collecting area; these are based on fundamental laws of physics involved that can't be broken – even by Takahashi or Tele Vue!

Writing Good Reviews

Writing a review of your astronomical equipment is a worthwhile endeavour for many different reasons. For one thing, being critical about your equipment encourages you to look carefully at (and through!) your equipment, comparing it to other pieces of equipment you own. Examining separation of double stars using different eyepieces, or noting changes in brightness of nebulae when shuffling between telescopes of different aperture, is a great way to appraise your equipment and determine the best combinations for different targets and observing conditions. These sorts of tests also help to develop observing skills. More generally, writing a good review is way of sharing your experience with others. The benefits of this are many: people shopping for similar equipment to your own can use your reports to make informed choices, and communicating your thoughts with other astronomers helps to develop the community feel of the hobby making everyone feel more involved. Serious reservations about the quality and performance of a product stated in a review that is widely read and from a respected writer may even encourage manufacturers to improve the quality of their products.

What is the best way to start writing a review? Unquestionably, the beginning is in deciding whether you are going to compare the telescope, eyepiece or whatever against the market in general, or to concentrate on the specific performance of the item using clearly stated criteria. As noted earlier, the first approach really only makes sense if you have a reasonable amount of experience and have actually used comparable astronomical equipment with which to draw your conclusions. In contrast, absolute statements on the performance of a piece of equipment are possible to make whether you are an old hand or a newcomer to the hobby. Compare the statement "this telescope gives the best view of Saturn I have ever seen" with "I could clearly see the Cassini Division of Saturn's rings and four of the moons". The former statement is a qualitative opinion, while the latter is more a quantitative fact. The advantages of the second way of describing something should be obvious: it doesn't depend on the users experience of other instruments and it doesn't take a view as to whether the results were good, bad or indifferent compared to other instruments. Instead, it simply gives the facts. Of course you will want to say whether or not you like the equipment and think it is performing well, but to start with it is always best just to record what it can and cannot do in terms of bare facts.

One popular debate among amateur astronomers is the value of the telescope star test. Some consider it an indispensable method for detecting and categorizing any flaws or imperfections in the lenses and mirrors of a telescope; others believe it a total waste of time. Appendix 1 includes some links to web sites about collimation, and these explain in detail how to carry out a star test and what to look for, but in short, looking through a telescope aimed at a star should reveal not a tiny point of light but a small disc surrounded by a few faint rings, even when focused properly. Slight defocusing of the image exaggerates this, making it easier to see the rings and the disc (called an Airy disc after George Biddell Airy, the British Astronomer Royal who made a study of the phenomenon). Essentially the more light in the central disc compared with the rings surrounding it, the better, and the shape of the disc and rings (they should be perfectly concentric circles) gives clues to any shortcomings in the optical system such as distortions in the mirror. The disk and rings should also look the same both when brought out of focus in both directions. If you take a telescope outside from a warm room and try to see the Airy disc, you will not have much luck though. Instead you'll see the rings boiling away around a blurry, spiky blob – what you're seeing is the movement of warm air currents in the optical tube refracting the beams of light. So for the star test to be meaningful the inside of the telescope must be at exactly the same temperature as the air outside, and this may take several hours to achieve depending on the design of the telescope (open tube reflectors and small refractors cool down fastest, while SCTs and Maksutovs take the longest). At the very least, understanding the star test helps when collimating a reflector or SCT by showing you what adjustments to the mirrors are required. However, if you have a closed-tube telescope without collimation screws, such as a refractor or a Maksutov, there's very little you can do if the star test doesn't come out perfect beyond contacting your dealer for an exchange or repair of the faulty unit if the flaws are serious.

Amateur astronomers are very keen on writing down their impressions of a piece of equipment on the first night of use. This is fair enough, first impression *do* count, but before extolling or condemning the equipment under review on the basis of that first night, take a few nights to put the thing through its paces. For a start, sky conditions such as seeing and transparency might not be perfect on that first night, and other factors like the altitude of your observing target above the horizon and ambient light pollution can come into play as well. However, there is also an element of practical competence to consider at well; it takes a while to get used to new piece of equipment and to fully understand how to use it. Therefore, it makes sense to spend some time just using your new tools and carrying out your usual observing program before feeling compelled to write your review. Indeed, there is a very good argument for not writing a review for several months after acquiring the thing you want to write about. That way you will be comfortable with the new equipment and will have used them under a wide range of different observing conditions and on a variety of targets. As with any field of technical writing, the more comprehensive your experience is, the more authoritative and objective your writing becomes.

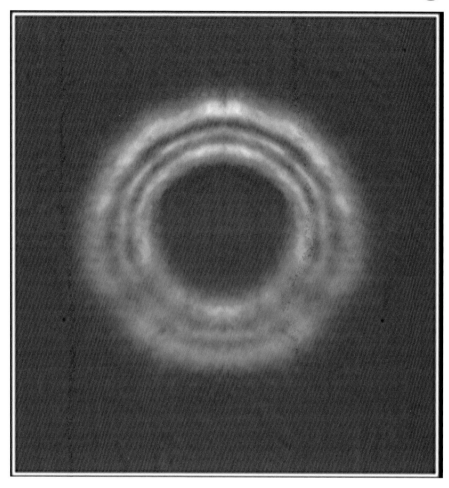

Figure 2.2. Provided the seeing is good and the telescope has reached a thermal equilibrium with the surrounding air, even casual examination of out-of-focus stars (in this case Vega) will show if a telescope needs collimation. Here, the bright diffraction rings form nice clear circles centered around the shadow of the secondary mirror. If the rings are elliptical, then collimation is required. Closer examination will reveal other flaws, such as poor optics; numerous web sites, magazine articles and books go into this subject in great depth (see Appendix 1).

Personal Astronomy-Based Web Pages

Writing equipment reviews and publishing them on review sites is one of the simplest ways to share your experiences and opinions with others in the hobby, but is not the only way. Many amateur astronomers have personal web pages that contain, among other things, reviews of their equipment. The extent to which these provide useful information varies of course depending on the range of equipment owned by the individual and their experience of other instruments and general depth of knowledge of the hobby, but done well these sites can provide valuable tips on particular items. Some of the authors of these sites can be passionate about their equipment, describing in lavish detail their observations and every possible adjustment to an instrument that will squeeze out the every bit of performance the thing can deliver. Others are much more laconic, with terse little summaries of each piece of kit, what they use it for, and whether or not they like it. Ultimately though, these reviews are included because they are among the easiest and most straightforward articles that can be written and posted onto a web site. Hobbyists that are more ambitious include things like astronomical photographs, educational resources such as how-to essays, calendars of forthcoming astronomical events, and so on. All these sorts of projects require planning to work really well, but they can tremendously useful to other hobbyists, especially newcomers to the field who really appreciate getting firsthand information on topics like the best way to observe deep sky objects or how to get into astrophotography on a budget.

This book isn't the place to talk about the intricacies of web design. There are many books on that subject, and the easiest thing is simply to visit web sites you really like and just copy them as best you can! Nevertheless, it is worth looking briefly at the fundamentals if only you spur you into having a go. The tools available are manifold and come suited to every pocket and for every computer operating system, but there are two main ways to go about. Some people like to create their web pages by writing out the HTML (HyperText Mark-Up Language) the old-fashioned way, by hand. HTML is the stuff the text part of web pages, including commands like link. A text editor a program that works a bit like a word processor but doesn't use font styles, formatting, etc., instead producing text documents of a very pure sort, called ASCII (American Standard Code for Information Interchange). There are text editors for every computer platform, such as *Notepad* for Windows; *BBEdit* for the Mac; and *vi* and *pico* for UNIX operating systems such as Linux. Even word processors can export files as ASCII text, usually called plain text in contrast to the formatted text files they normally create. ASCII text can be read by any computer and in fact is much older than home computers as we know them now, having been devised in the early 1960s for teleprinter, or Telex, machines. Because ASCII text is universal, it makes an ideal medium for creating the HTML code to be stored on the server computer and read by the remote computer. Writing HTML by hand in a text editor is effective but time consuming, and frankly, more than a bit heavy going. HTML isn't a particularly complicated language to learn compared with proper computer programming languages or even scripting languages like Java or AppleScript, but it is full of seemingly arcane symbols and phrases nevertheless. So many people, particular amateur web page developers, prefer to use "web-authoring" packages with a more intuitive, usually word processor-like interface. Indeed some word processors like *Microsoft Word* and *AppleWorks* (both of which are available for Windows and Macintosh computers) produce nice if relatively simple web pages. Linux users can look to Sun's *StarOffice* or the open source *OpenOffice* for the same sort of functionality. One version of the popular Windows and Macintosh web browsers, *Netscape Communicator*, even includes a decent web-authoring package, and better yet is free! Nonetheless, the dedicated web-authoring tools are more versatile and powerful, and usually easier to use. These range from consumer-level applications like *Microsoft FrontPage* (Windows) and *Adobe GoLive* (Windows and Mac) through to the high-end (and much more expensive) heavyweights *Macromedia Dreamweaver* and *Adobe LiveMotion* (both available for Windows and Mac). Aimed primarily at professional web designers, these top-of-the-line programs give even the greenest amateur the ability to include things like rollover buttons, frames and cascading style sheets into their web sites easily. These things are painfully slow to create with text editors and usually impossible to do with the minimalist web-authoring tools in word processors.

Once you have the tools to build your web site, the next thing is to decide on the content to put on the pages. Since our theme here is astronomy, your job is to figure out what it is you have to say, and how it differs from the thousands of other astronomy sites on the web. This latter point is crucial if you want visitors to stop by in appreciable numbers. Therefore, you might decide to slant the content towards observers of a certain type of object, perhaps the planets or variable stars. Alternatively, you might build your site around a particular piece of equipment or tool, to help inform other owners of that equipment on your experiences, problems and successes. These sorts of sites are always very popular because they are so specific; books tend to cover things in generalities to maximize their appeal, but a web site can be much more focused on a certain tool or process. Another way to make a site attractive to visitors is to aim at certain types of amateur astronomer, perhaps young children, newcomers to the hobby, or those with physical disabilities that makes some of the commonplace methods difficult or impractical. Working in this way you can create essays and projects that lead the visitor through the hobby helping them develop their skills. A final example of a good approach to take is to create a community around it, perhaps by making it the home page for your astronomy club or the science department at your school. Different people can contribute descriptions of their observations of a particular night, reviews of the equipment used, and announcements of events and star parties.

The actual content itself will depend on your web authoring skills, the tools you have and the amount of time and effort you are prepared to expend. Simple web pages might include mostly text with a few pictures of your equipment, taken using a digital camera or webcam. Both these instruments produce digital images that are easy to incorporate into web pages, either directly or by converting into a more web-friendly format. Digital cameras commonly produce still images using the JPEG (or Joint Photographic Expert Group) compression method that display nicely on any graphics-aware web browser, so are suitable for uploading to your web site pretty much as they are. Pictures and movies taken from webcams are a little more variable though, depending not just on the webcam used but also on the operating system of your computer. A certain webcam used with a PC might produce AVI (or Audio Visual Interleave) movies but QuickTime movies when used with a Mac. Both of these popular formats play on most web browsers, but sometimes extra bits of software (called plug-ins) are required. The great advantage of webcams over digital cameras as far as digital astrophotography is concerned comes not from the movies themselves, but the ability of certain programs to extract individual frames and then *stack* them together to make brighter, sharper composite images. Among these programs are *AstroStack* (for Windows) and *Keith's Image Stacker* (for the Mac), and Chapter 6 will go into this sort of digital astrophotography in depth.

More challenging would be web pages including instructional star charts showing visitors how to find deep sky objects. If done well these are much appreciated by beginners, who often find it difficult to find the celestial sites more experienced observers take for granted. Planetarium programs, such as *TheSky* and *Starry Night*, can create exportable star charts, and usually these charts are *bitmap* files of some sort, normally JPEGs. Bitmaps are pixelated images that look good on screen but being relatively low resolution (72 dots per inch) doesn't look all that nice when printed out. Fine detail and small font size text can be very difficult to read, so make them as big as is practical and avoid using a small typeface for the labels. Bitmap star charts are best where you want the user to study them on the computer screen, perhaps along with some instructional text explaining how to make the star hop from an obvious bright star to the deep sky objects in question. If you want your readers to print off the star charts and use them at the telescope, then you might do better avoiding bitmaps and using *vector* images instead. Vector images are not mosaics of pixels like bitmaps, but precise curves and lines that can be scaled up or down without text or details becoming obscured. PostScript files the best known of these sorts of files. The high-end planetarium programs *SkyMap Pro* and *XEphem* are examples of programs that can produce publication-quality PostScript star charts. We'll be looking deeper into the differences between these two different formats for producing star charts in Chapter 3, but for now it is well to be aware of the differences.

Images and pictures are a central part of making web pages look good, but words are just as important because they give a web site depth. We've already mentioned equipment reviews but what other sorts of things are worth writing about? Observing reports are popular and useful because they help others understand how to use their telescopes and what objects to look for with a telescope of a given size, and what things look like for real as opposed to the pretty pictures shown in most books. Describe the telescope and eyepieces used, the targets, and so on. Don't forget to mention the viewing conditions: whether or not the Moon was up; how steady the air was (the *seeing*); and how clear it was (the *transparency*); and how dark the skies were (the ambient *light pollution*). Appendix 2 includes some information on how to judge these criteria and so make your report as valuable and objective as possible. Before we leave the subject of sky conditions and observing, suburban astronomers need not despair about the fact that for deep sky observing dark skies are the best. The Sun, Moon and planets yield just as much detail under light polluted skies as they do out where the skies are beautifully dark. Many of the astronomy magazines and web sites ignore city-bound astronomers, and this would be a great topic for an amateur seeking to create a web site offering something new.

An Example of an Observing Report

The following is an example of an observing report using the sorts of scales described above, and formatted for inclusion in a web site. I've tried to give the thing some sort of narrative, so that the reader can build up a picture of what I did, and some tips on viewing the objects that I found useful. Always keep in mind the audience: terse, dry text is fine in small doses but causal readers may lose interest if there's too much. Be friendly, explain why these things are worth looking at, and try to write for readers of all different ages. Most important, don't let the science get in the way, or try to impress the reader with how much you spent on your eyepiece. Instead, share your enthusiasm with the reader in as easy and contagious way as you can.

Date and time of Observation: April 10/11 2001, 10 pm – Midnight
Observing Location: Hertfordshire, England
Objects Observed: Some nice springtime doubles and NGC 6543 (Cat's Eye Nebula)
Seeing: IV
Transparency: 5
Sky Darkness: Bortle Class 5 (Suburban sky)
Moon: Waning gibbous (16 days), above horizon
Telescope: Celestron Firstscope 114 Premium (4.5 inch Newtonian)
Eyepieces: 20-mm Plossl, 10-mm SMA, ×2 Barlow

Observing notes: Started off in Cancer, to see one of the treasures of the early spring skies, Iota Cancri. This is a lot like Albireo, consisting of a blue star and an orange star. Then, stepped back along the ecliptic to Leo. Gamma Leonis (Algeiba) and 54 Leonis are two nice double stars. Even at low (×45) power Gamma Leonis splits easily to reveal one yellow and one orange star, while 54 Leonis may need medium power (×90) to be separated into its blue-white and steel colored components. Although subtle, this is a very attractive double star.

North of Leo is the constellation Lynx, a string of faint stars that is easily overlooked. This is a shame, as it contains some really nice objects, including some lovely multiple stars. 38 Lyncis is the not very far from the sickle of Leo, and quite tricky in very small telescopes, but a well collimated 114-mm reflector should have no problems. At the other end of the constellation, near Capella, are 12 and 19 Lyncis. These are triple stars, 19 Lyncis being the more open of the two, but otherwise quite similar. Each has two bright stars close together and one fainter and more distant companion. Stacking the 10-mm SMA on the Barlow to get a high power (×180) is needed to split these two triple stars.

Adjacent to Lynx is Ursa Major, which has one of the most famous double stars, Alcor and Mizar. It's a naked eye double, and a good check not only on your eyesight but also on how well dark-adapted you are. Through the telescope, Mizar is revealed to have yet another companion. Ursa Major adjoins two fainter constellations that contain some double star treasures: Ursa Minor and Draco. At the tip of the Little Bear's tail is Polaris, that is a not an easy split for a 114-mm telescope. Although the companion to the primary is 17 arc-seconds away, it is very much fainter. The trick is to use high power (×180), and in reasonably steady skies and dark-adapted eyes, the eighth magnitude companion is clear enough. At the other end of the Little Bear is Gamma Ursae Minoris, an easy blue and orange double star. The lowest power is needed here, and in fact the finderscope or binoculars give better views! 16 and 17 Draconis, near the edge of Draco and not far from Hercules, are another easy double for binoculars. My "double star trek" in this part of the sky finished off with 39 Draconis. It's a triple star, which needs a well-colli-mated telescope and quite steady skies to split all three stars (though A and C can be seen in binoculars).

While in this neck of the woods, I tracked down NGC 6543, the Cat's Eye Nebula. Through a 114-mm reflector, this is obviously a planetary nebula – a large and rather bright disc. As it happens, NGC 6543 is very close to another star, and at low powers (×45) the two fit easily in the field of view. Focus on the neig-boring star to get the best view of the nebula. There isn't much to be see; the fun is in tracking it down, it is much more difficult to find than the Ring Nebula (M57) but to me at least seems brighter and bigger.

Mailing Lists and Newsgroups

Web pages might be the simplest way to show off photographs or observations or share tips and ideas, but they are a one-way method of communication. Admittedly a visitor can e-mail the author of a web site to ask for more or pass on a comment, but for an interactive discussion with lots of people with similar interests a more dynamic such as *mailing lists* works much better. Mailing lists forward messages sent by one subscriber to the mailing list to all the other sub-scribers of that list. If they want to, they can then reply, and everyone else will be able to read that reply. A mailing list is a closed, or private, forum, that is, only subscribers can see messages or reply to them. Such lists are especially useful to beginners (or *newbies* in Web-speak), giving them the chance to ask questions in a forum where the expertise to answer them is in good supply. Some subscribers reply to questions frequently, and may even compile documents called FAQs (or Frequently Asked Questions) to help the newbies. Other subscribers rarely ever reply, instead preferring to read the messages and learn from them; these individ-uals are the *lurkers*. All subscribers are expected to stick to a few rules of manners and politeness, known as *netiquette*, but otherwise these groups are generally friendly, informal and very useful.

Halfway between a web site and mailing list are *newsgroups*. Although these are based around text messages, just like mailing lists, they aren't sent directly to your e-mail account. Instead, messages go to a newsgroup message board where anyone can read and reply to them, not just a limited group of sub-scribers. Some Internet service providers offer direct access to newsgroups via software such as *Microsoft Entourage* or *Netscape Communicator*, but others don't. In that case, web sites such as Google provide an adequate, if slower and less configurable, alternative. There are several astronomy-based newsgroups of which *sci.astro.amateur* is perhaps the most vigorous; not to mention dozens of newsgroups devoted to different aspects of computer hardware and soft-ware. Although accessing newsgroups is a bit more cumbersome than mailing lists, they do have the advantage of a much wider audience. A downside to this is the tendency for companies that send out *spam* – unsolicited commercial e-mail – to use addresses they find on newsgroups. Many users of news-groups like to camouflage their e-mail addresses to prevent this (such as j.smith@_REMOVE_THIS_BIT_myserver.com). It is easy enough for a human being reading this e-mail address to see what needs to be altered to make it work properly, but for the purposes of spam e-mail this address is useless, since spammers rely on computer programs that copy addresses blindly without checking to see if they make sense or have been altered in some way.

Asking Sensible Questions

There is an old computing proverb that (rewritten politely!) says "nonsense in, nonsense out" and what this means is that you will only get a useful answer if you begin with decent data. This is very true with newsgroups and mailing lists: the more specific and detailed your question is, the more accurate and reliable the answers you get are likely to be. Add information that might be relevant for example your problem is getting a go-to telescope to align, state not just the type and model but also details like the firmware used by the telescope's handset and the software running on your laptop. Similarly, if you want to know about the best piece of equipment to do a certain task, the additional criteria that will influence your choice, such as budget, the ages of any children likely to be using the equipment, and so on. Computer-related questions are among the most difficult to answer because different users of a certain program will often have very diverse set-ups. So, if you are asking about a program that doesn't work properly, make sure to include things like the computer model and operating system, the amount of memory installed, processor speed and the version number of the software in question. Netiquette is important, too. This means avoiding the use of capital letters, for example. These are not only difficult to read, but when they are used, are meant to represent shouting. Be polite, and say thank-you to the people who answer. Often, follow-ups to questions are valued if shared with the list or newsgroup in general. So, if the recommendations of a certain person helped you to solve a problem, let everyone else know so they can benefit from your experience.

Giving Useful Answers

One of the wonderful things about any hobby is that after a while you find yourself giving advice to others; in short, you've become an expert. When that happens, you'll soon realize how much you've learned from others, and that many of the things you've mastered are actually quite difficult and not at all intuitive. Amateur astronomy is like this because it is a practical skill as well a science: it doesn't matter how much reading you do, it is the time spent locating objects in the night sky and then observing them at the eyepiece that makes the difference. All the same, getting experience second-hand by asking for advice from others is useful and one of the best uses of the Internet.

Obviously giving good advice depends on you having the experience and knowledge to solve the problem, but it also helps if you care to explain and articulate things well. In a technical and intrinsically difficult subject like astronomy,

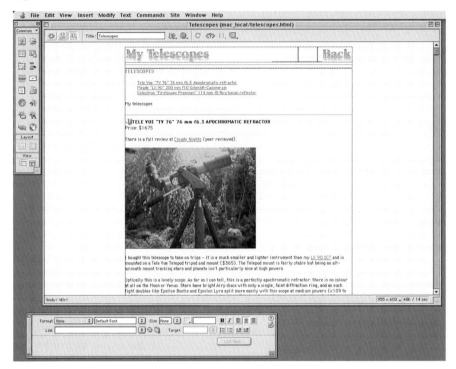

Figure 2.3. Web-authoring packages, such as *Dreamweaver*, make creating professional-looking web pages relatively simple.

we're in this strange situation of exploring an aspect of science beyond that which many people studied at school or college. Professional teachers often visualize their work as building bridges between what the student *already knows* and what they *need to know*. Likewise, if the person you are trying to help is clearly very new to the subject, work out their level of experience and understanding, and then build your bridges from that. The thing *not* to do is to bamboozle the questioner with jargon and abbreviations that might mean something to you but leave them completely in the dark. However impressive *you* might think it makes you sound, it won't help *them* in the least. Master explaining things, and you're well on your way to be becoming an ambassador for this wonderful hobby!

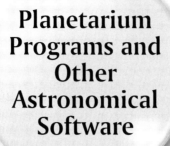

CHAPTER THREE

Planetarium Programs and Other Astronomical Software

There is huge amount of software available for amateur astronomers, ranging from the ubiquitous planetarium program through to specialized tools like image processing software and utilities for controlling telescopes. In addition, not-specifically astronomical software has been pressed into service as well, for example, databases for storing information on deep sky objects and spreadsheets for calculating magnifications of eyepieces in a given telescope. All this is before you even start looking at the software designed for research and educational purposes, and online resources like the images that make up the Digitized Sky Survey (a professional project comprising photographs of the entire night sky at great detail using British and American telescopes). In this chapter, we will be taking a look at the range of software available and the sorts of things that they do.

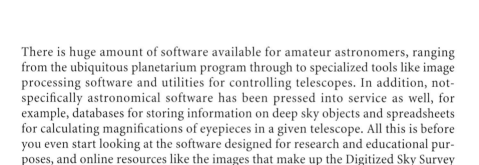

Using Software in the Field

Whatever software you choose to buy and use, an important issue is how best to use it alongside your telescope. If you plan to use the software indoors on a desktop computer, then this topic is not relevant at all, but if you are going to use the software on a laptop at night, then you need to be aware of the fact that a bright laptop screen is a very undesirable thing to have around. Laptop screens will ruin your *dark adaptation*, making it difficult, even impossible, to see faint deep sky objects through the telescope.

Dark adaptation is the ability of our eyes to become more sensitive to light than is normally the case by day. Humans have fairly good night vision when fully adapted to the dark, though inferior to that of most other mammals. What our eyes do well though is provide high-resolution, color imaging far superior in those regards to what your cat or dog sees by day or night. This is because our eyes contain large numbers of cone cells that are sensitive to colors, one kind to red, a second to green and a third to blue. Each cone cell has its own personal connection to the optic nerve, so each one operates like a tiny pixel on a CCD, allowing the brain to produce a high-definition image. The downside to cones is that they do not work at all in the dark. When it gets dark, we rely on cells called rods instead; these work under low light intensities but are completely insensitive to color, detecting only the presence or absence of light and not its wavelength. Moreover, rods do not work independently but in groups, sharing a connection to optic nerve. While this increases the chance of a dim light triggering the optic nerve into sending a message to the brain, it makes it impossible for the brain to know precisely which rod was stimulated. This means that rods produce blurrier images than the cones since each group of rods acts like a much coarser "pixel" than the cones. However, since the central part of our retina contains just cones, the middle part of our view is sharp, and this is the bit we use for looking at things, and that allows us to see small details clearly such as the words on this page. Most other mammals have retinas that contain only rods, so they see the world in a blurry, black and white way compared to us, even if they can see things more brightly in the dark.

All this biology has some important implications for astronomers. For one thing, in means that as the Sun goes down and it gets dark, the cones that we rely on by day become less useful. The Moon and planets may cast enough light through the telescope to trigger them, as do many stars, but deep sky objects in particular are far too dim and so can only be perceived in shades of grey: a clear sign that it's the rods and not the cones that are sending the signals to the brain. As mentioned before, the cones are in the central part of the retina, so if you want to cast the image on those cells that work best under low light levels, the rods, you need to look off to one side. This is the *averted vision* technique frequently mentioned by advanced astronomers. When the image is on the peripheral part of the retina, it can be quite surprising how much brighter it seems. To give you an example of this effect, even under fairly dark rural skies through a 200-mm (8-inch) SCT, good globular clusters like M13 and M3 still look like fuzzy blobs to me when viewed straight on, with just a few stars around the edges actually being obvious. With averted vision though, these objects really do start to look like the photographs, if not actually bright, then certainly resolved pretty much to the core. As our eyes adapt to the darkness, which means not only that the pupil gets bigger but that the rods themselves become more sensitive to dim light, this effect gets even more pronounced. One important thing about night adaptation is that is much easier to lose than develop. To properly become dark adapted, you need to be in the dark for at least thirty minutes, but reversion to bright light conditions, known as *light adaptation*, takes just seconds. This is why flashlights and other bright artificial light sources are so unpopular around amateur astronomers – they ruin the critical dark adaptation that takes so long to develop. This rapid light adaptation isn't a bad thing, it's essential: without it, a sudden bright light would damage the retina, whereas the opposite, exposing a light adapted eye to darkness, is inconvenient to astronomers perhaps but doesn't cause any damage at all. That our eyes are able to work across such a range of brightnesses is really

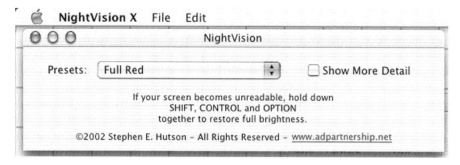

Figure 3.1. *NightVision* is a utility for Windows and Macintosh computers that darkens and reddens laptop screens, and so helps to protect the all-important dark adaptation vital to proper observing of deep sky objects and faint stars.

quite a remarkable fact, and this "dynamic range" is far beyond anything possible with man-made cameras.

Obviously turning a laptop screen down to its dimmest setting will help, but to protect your night vision when using a computer you need to change the screen's color as well. Normally the computer uses a full color palette to produce colored images made up of red, green and blue pixels. Our eyes react most strongly to green light than to red, and that's why if we see color in deep sky objects like the Orion Nebula, they seem green rather than red even though photographs show the opposite. Many planetarium programs take advantage of this sensitivity to different colors, by not just dimming the screen but giving it a red cast as well. This means the screen can be a little bit brighter, and therefore clearer, than if it was working in full color. If your planetarium program doesn't have a night vision option, then there are alternatives. Turning down the brightness is one option but it really isn't all the effective because even at the lowest setting the screen will still be far too bright. Windows and Linux users can change some of the user interface settings from the normal colors over to shades of red, and this will help a bit more. But the best option then is to download a night vision utility like A D Partnership's *NightVision* or Ilanga's *NightMaster* (see Appendix 1 for the web addresses) that will covert the display to a night vision palette on a system wide basis. Either of these will allow you to manually switch the display to a night vision mode. In fact, even if you don't use a planetarium program these can be useful if you use your laptop to record observations, collect images from a CCD or webcam, or even just use your computer to play MP3s while observing!

Planetarium Programs

Planetarium (or sky charting) programs are by far the most popular type of astronomical application. Some are comparatively basic, sticking to the core function of mapping the night sky onto a computer screen, usually in a configurable way that allows the user to find objects, zoom into particular fields more closely, or toggle on or off certain types of object. Others are much more complex, and offer features like the ability to control go-to telescopes and are able to connect to online information sources such as the Space.com and the Digitized Sky Survey. Some can even help plan observing sessions by identifying the best times to view showpiece objects as the night progresses. In price, the range from freeware through moderately priced shareware up to expensive commercial applications costing about the same as a beginner's telescope or a deluxe eyepiece. Here we'll examine the range of different programs available at the time of writing (2003), and though software does change, to be honest the line-up of planetarium programs has been relatively stable; many programs being updated every year or two. Another interesting aspect of the range of planetarium programs is that despite the large number of products all seemingly designed to do much the same thing, most of them are very distinctive and deal with the star-charting software paradigm in dramatically different ways. Take for example the way programs allow the user to magnify regions of the sky simulation. Some, such as *Starry Night*, treat the sky simulation like a flat two-dimensional map and use a magnifying glass metaphor (complete with magnifying glass cursor!) for zooming in. In contrast *The Digital Universe* treats the simulation as part of a globe, admittedly flattened out onto the screen, and the cursor is used to marquee off quadrilateral regions with obviously curved edges that are then re-drawn, and pulled into straight edged rectangles, to occupy the full screen.

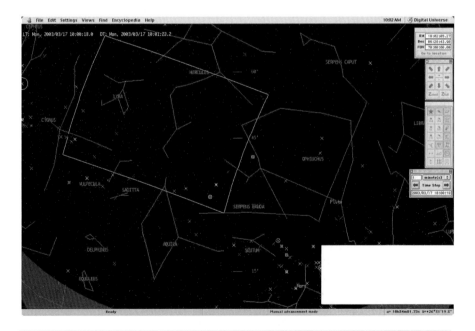

Figure 3.2. *The Digital Universe* uses a globe-like paradigm for its projection; although a bit odd to use at first, it does fit our perception of the night sky rather well.

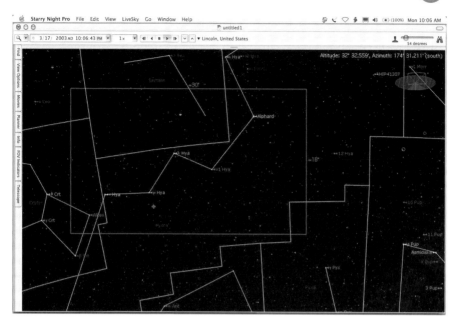

Figure 3.3. *Starry Night* is typical of majority of planetarium programs using a map-like projection of the sky that is certainly an intuitive approach but not always easy to fir with hemispherical dome of the night sky we see when in the field.

Because the cost of a program is very often a key factor in purchasing decisions, I think it is important to consider the cost of a program when balancing its feature set against those of another program. Actually quoting the price is pretty irrelevant, they change from year to year and from location to location; instead I'm going to divide the programs into three price groups: those that are free, those that are budget priced, and those that cost serious money. Free software includes stuff produced by altruistic amateurs, like the popular Windows application *Cartes du Ciel* and the open source *KStars* for Linux. Budget software, for our purposes, is the stuff that costs around $20–50, or about the same a Plössl eyepiece. Most of this software is shareware, such as the Windows program *DeepSky 2003* and *Equinox* for the Mac; but there are some low-cost commercial software available in retail stores as a minimally packaged or jewel case products that cost about the same as a generic music CD. Often these sorts of programs are relatively old though, so their apparent bargain-price can be a bit misleading; it's easy to find shareware titles that cost the same but have a much better feature set. Unlike retail software, you can try out shareware before purchasing. Sometimes the software is time limited or has certain functions disabled until it is paid for, but during the trial period you'll pretty soon decide whether the application suits your needs and so decide whether to keep the program and pay the developer for it. A key advantage of the shareware business model is that it is viable even with relatively small markets because the costs of production, promotion and distribution are so low. This means that while the commercial producers concentrate primarily on Windows users because they account for most of the retail market, you can find good, modern shareware software available for users of all operating systems. The final category of software as far as price is concerned is for the high-end commercial programs like *TheSky* and *SkyMap Pro* that cost around $150, sometimes less but often much more if the program comes with power-user features like go-to telescope control and comprehensive deep sky databases. If the budget stuff costs what a cheap Plössl does, then these programs come in nearer to what you'd spend on a Nagler, and the people who buy these applications are just as likely to own a set of premium eyepieces too. One thing about planetarium programs is that they don't age quickly, and can be a real one-off purchase, so their expense is less of an issue when looked at over the long term. Because you don't need to swap files with other users like word processors, there isn't the constant scramble to maintain compatibility, and the stars and galaxies don't change their positions quite as often as Microsoft updates its software!

Choosing Windows, Mac or Linux Planetarium Programs

Surprisingly perhaps, cost alone isn't a good indicator of the capabilities of a program. Some of the most powerful applications are free, while there are some commercial programs especially designed to be simple and easy to use. To get the best value from a planetarium program, you want one suited to your needs. Planetarium programs fall into three categories: lightweight ones primarily for families and absolute beginners; middleweight programs adequate to the needs of most casual amateur astronomers; and finally the heavyweight programs with the super-sized catalogs for advanced observers using very large telescopes.

Table 1 outlines the key differences between these three groupings. The first three rows should be self-explanatory. The fourth, "Other Deep Sky Catalogs, is somewhat vague but this refers to whether or not a program has catalogs that go beyond the thousands of objects on the NGC and IC lists, for example the Perek and Kohoutek Catalog of planetary nebulae and the Abell Catalog of rich galaxy clusters. What it doesn't mean are new lists made up of choice pickings from the NGC or IC, although such lists as a way to wean new observers off the Messier objects can be very useful indeed! The next row is for whether or not a program allows for "User Expandable Databases, such as catalogs not included in the software as installed but downloadable from the Internet. The "Solar System Updates" refers to things like new comet and asteroid databases and the position of the Great Red Spot on the surface of the Jupiter. The next three categories refer to various moons in our Solar System. The first of these is a program plots the positions of the satellites of Jupiter and Saturn. This may be by showing them directly on the sky simulation or as elongation charts similar to those published in astronomy magazines. At the very least, the four bright moons of Jupiter (the Galilean moons) and Saturn's brightest moon Titan should be plotted since these are the ones that can be seen even with a small telescope. More advanced astronomers with 100 mm (4-inch) or larger telescopes will easily see several

Table 3.1. A comparison of the features found in lightweight, middleweight and heavyweight planetarium programs

	Lightweight	Middleweight	Heavyweight
Stars	100,000	1–2 million	Several million
Messier Catalogue	Yes	Yes	Yes
NGC & IC Catalogues	No	Yes	Yes
Other Deep Sky Catalogues	No	Maybe	Yes
User Expandable Catalogues	No	Maybe	Yes
Solar System Updates	No	Maybe	Yes
Jovian & Saturnian Moons	No	Yes	Yes
Martian & Uranian Moons	No	No	Yes
Lunar Ephemeris	No	Maybe	Yes
Observation Planning Tools	No	Maybe	Yes
Image Manipulation	No	No	Yes
CCD Integration	No	No	Yes
DSS Internet Linking	No	No	Yes
FITS Image Viewer	No	No	Yes
Telescope Control	No	Maybe	Yes
Field Of View Indicators	Maybe	Yes	Yes
Night Vision Mode	Maybe	Yes	Yes
Internet Resource Linking	No	Yes	Yes
PostScript Printing	No	No	Yes

more moons around Saturn, and may want these shown as well. The moons of Mars and Uranus are much fainter and much more difficult to see than the brighter moons of the two gas giants, and not many astronomers ever get to see them, but some programs do offer plots of their positions. Our own Moon is of course a popular target for observers of all skill levels and with any instrument. Most applications provide things like the Moon's position in the sky and its phase, but here the "Lunar Ephemeris" entry is for more advanced tools such as showing the degree of libration and its direction, and for features such as the identification of the craters and seas. Moving on to more practical issues "Observation Planning Tools" come in various forms but all serve to optimize your observing sessions by timetabling the deep sky targets and astronomical events to best effect, such as for when they are highest above the horizon. "Image Manipulation Tools" are those designed to make putting new photographs of deep sky objects onto the sky simulation, for example allowing the user to align the stars in the photograph of a star cluster with those on the simulation, and thereby getting the scale and orientation of the two to match. Some planetarium programs go further including not just image manipulation tools but also "CCD Integration", that is software tools for controlling CCD cameras and thereby providing features like focusing, imaging and calibration. For those content to view the labours of others, "DSS Internet Linking" allows the user to access the Digitized Sky Survey via the Internet from within the planetarium program, normally by selecting an object on the sky simulation and issuing the appropriate command. A "FITS Image Viewer" allows the user to view a versatile file format widely used by professional astronomers. Admittedly, not many amateurs will

Table 3.2. A comparison of the features found in twenty popular planetarium programs for Windows, Mac and Linux computers

	Stellarium	SkyGazer	The Sky (Student)	SN Beginners	Stargazer's	Touring...
Platform	Windows, Mac & Linux	Mac	Windows & Mac	Windows & Mac	Mac	Windows
Stars	1–2 million	100,000	100,000	100,000	100,000	100,000
Messier Catalogue	Yes	Yes	Yes	Yes	Yes	Yes
NGC & IC Catalogues	No	No	No	No	No	No
Other Deep Sky Catalogues	No	No	No	No	No	No
User Expandable Catalogues	No	No	No	No	No	No
Solar System Updates	No	No	No	No	No	No
Jovian & Saturnian Moons	No	Yes	Yes	Yes	Yes	No
Martian & Uranian Moons	No	No	No	No	No	No
Lunar Ephemeris	No	No	No	No	No	No
Observation Planning Tools	No	No	No	No	No	No
Image Manipulation	No	No	No	No	No	No
CCD Integration	No	No	No	No	No	No
DSS Internet Linking	No	No	No	No	No	No
FITS Image Viewer	No	No	No	No	No	No
Telescope Control	No	No	No	No	No	No
Field Of View Indicators	No	No	No	No	Yes	Yes
Night Vision Mode	No	No	No	No	No	Yes
Internet Resource Linking	No	No	No	Yes	No	No
PostScript Printing	No	No	No	No	No	No

Table 3.2. A comparison of the features found in twenty popular planetarium programs for Windows, Mac and Linux computers (continued)

RedShift	Alpha Centaure	SN Backyard	Voyager	KStars	Equinox	Digital Universe	SkyMap
Windows & Mac	Windows	Windows & Mac	Mac	Linux & Mac	Mac	Mac	Windows
100,000	1–2 million	1–2 million	Several million	1–2 million	1–2 million	Several million	Several million
Yes	Yes	Yes	Yes	Yes	Yes	Yes	Yes
No	Yes	No	Yes	Yes	No	Yes	Yes
No	Yes	No	Yes	No	No	Yes	Yes
No	Yes	No	Yes	No	Yes	Yes	Yes
No	Yes	Yes	No	No	No	No	Yes
Yes	No	Yes	No	No	Yes	Yes	Yes
Yes	No	Yes	No	No	Yes	Yes	Yes
Yes	No	No	No	No	No	No	No
No	No	Yes	No	No	No	No	Yes
No	No	No	No	No	No	No	No
No	No	No	No	No	No	No	No
No	No	No	No	Yes	No	No	No
No	No	No	No	No	No	No	No
No	No	No	No	No	Yes	No	Yes
No	No	No	Yes	Yes	Yes	Yes	Yes
No	Yes	No	Yes	Yes	Yes	Yes	Yes
Yes	No	Yes	No	Yes	Yes	Yes	No
No	No	No	No	Yes	No	No	No

Table 3.2. A comparison of the features found in twenty popular planetarium programs for Windows, Mac and Linux computers *(continued)*

Cartes	The Sky (IV)	SN Pro	DeepSky	Megastar	XEphem
Windows	Windows & Mac	Windows & Mac	Windows	Windows	Windows, Mac & Linux
Several million	Several million	Several million	Several million	Several million	Several million
Yes	Yes	Yes	Yes	Yes	Yes
Yes	Yes	Yes	Yes	Yes	Yes
Yes	Yes	Yes	Yes	Yes	Yes
Yes	Yes	Yes	Yes	Yes	Yes
Yes	Yes	Yes	Yes	Yes	Yes
Yes	Yes	Yes	Yes	Yes	Yes
Yes	Yes	Yes	Yes	Yes	Yes
No	No	Yes	No	No	Yes
No	No	Yes	Yes	No	No
No	Yes	No	Yes	Yes	Yes
No	Yes	No	No	Yes	No
Yes	No	Yes	Yes	Yes	Yes
No	No	No	No	Yes	Yes
Yes	Yes	Yes	Yes	Yes	Yes
Yes	Yes	Yes	Yes	Yes	Yes
Yes	Yes	Yes	Yes	Yes	No
Yes	No	Yes	Yes	Yes	Yes
No	No	No	No	No	Yes

choose FITS over JPEGs for gracing their web pages, but if you're interest leans as much into the academic side of astronomy as the purely entertaining, then this feature will be useful. "Telescope Control" is clearly only useful if you own a go-to telescope, and means that the program can be connected to the telescope and used to issue commands as opposed to the usual use of the telescope handset. "Field Of View Indicators" are useful with any kind of telescope, even binoculars, allowing the user to mark off regions of the sky simulation corresponding to the field of view seen at the eyepiece. For star hopping this feature is indispensable. Another essential feature if you plan on using the laptop at night alongside the telescope is the "Night Vision Mode, whereby the screen is dimmed and given a red cast (usually), thus reducing the effect bright computer screens have of ruining night-adapted vision. Some planetarium programs offer "Internet Resource Linking" of one sort or another allowing the user to tap into online astronomy services to get space-related news, alerts of interesting phenomena, information and pictures on objects, and so on. Depending on how this is implemented this can be very useful to children and educators especially, although most amateurs will find at least some of these resources interesting. Finally, "PostScript Printing" is the ability to produce high-quality star charts based on vector graphics rather than bitmaps. If you want print quality graphics for a magazine or newsletter, this is essential.

By my reckoning, lightweight applications are those that have five or fewer features, middleweight ones less than ten, and the heavyweights ten or more. Having established my criteria for dividing the planetarium software market into these three levels, the following is a survey of these different programs. My hope is that this will help you make an informed decision when it comes to laying down the cash at your friendly neighborhood astronomy store. Nevertheless, these are my criteria, and not yours, although I have tried to include the ones most fundamental to backyard astronomy. Getting the right planetarium program for your needs is a vital part of enjoying digital astronomy: after all, these programs are pretty much the heart of any amateur astronomer's software collection, and the best of them can handle many of the tasks that you will want your computer to do.

Lightweight Planetarium Programs for Beginners and Children

Just because an application is "lightweight" doesn't mean it's a weak or bad application, but rather it concentrates on a narrow range of tasks, the ones most beginners are going to be most interested in, and does them with the minimum of fuss or the need for computing expertise. If you think about, most of the time we use any planetarium program to do a small number of things most frequently: identify the less obvious stars and constellations, locate deep sky objects or faint solar system objects like asteroids, and produce useful star charts to help us find these objects in the sky. Most of us don't really need a star catalog that with millions of very faint stars and deep sky objects we can't even see from our light polluted suburban gardens and porches, and so spending money to get this facility is largely a waste. Similarly, for observers with a small telescope, i.e., something less than a 200-mm (8-inch) aperture, even under pristine skies there isn't much point to finding some fourteenth magnitude planetary nebula on the computer screen when you haven't a hope of seeing it with your telescope. If you have a laptop that is no longer at the cutting edge of performance, then one of these modest programs could be very useful, since they usually require less processor power, memory and disk space than the bigger planetarium programs. These programs are also much simpler with fewer unnecessary bells and whistles, making them a great choice for children or those who just don't like fussy computer programs. In general then, if you can live without the massive star and deep sky catalogs that come with the middle and heavyweight applications, there's a lot to be said for this sort of software, and fortunately there are some nicely made, good value planetarium programs to consider.

So, what should be on your shopping list if you're looking to buy (or download) a lightweight but useful planetarium program? Top of my list would be the full Messier Catalog of deep sky objects together with at least the pick of the New General Catalog (NGC) and Index Catalog (IC) objects as well. None of us needs a planetarium program to find the Moon, and the brighter planets should be obvious too (if not, the weather columns of newspapers usually give this information). But finding interesting deep sky objects is often difficult for beginners and children because the sky outside your house never quite looks like the star charts and maps in books, yet this is crucial if interest is to be maintained, as deep sky objects are the most diverse and fascinating things amateurs can look at with their telescopes. The hundred or so Messier Catalog objects include some of the best, at least for northern hemisphere observers, and there are usually a handful up every night that are easy enough to find if you know where to look. Despite what many writers say about so many of these being "just visible to the naked eye" or "obvious through the finderscope" the sad fact is that under suburban or even semi-rural skies this just isn't so, at least, not to people starting out in the hobby. So having a star chart that can be tailor-made to your locality and time of night is a huge help in finding these sometimes elusive treasures. We'll look at exactly how you use planetarium programs to this end a little further in this chapter, but the key thing is that they can label and color stars, helping the user identify to the constellations. Most constellations are nothing like as obvious as they seem in books, particularly when they're upside down or half hidden by trees. Although the Messier objects include many of the best observing targets for northern hemisphere observers, it doesn't include them all, and misses most of the good southern hemisphere ones. Other lists include those, such as the NGC and IC. Admittedly, the bulk of both the NGC and IC catalogs include some pretty poor fare as far as observers with binoculars and small telescopes are concerned, but there are some showpiece objects that can be enjoyed with telescopes of any size, even the naked eye. The Perseus Double Cluster (NGC 868 and 884) is one of the most spectacular omissions from the Messier Catalog for northern observers, while southern hemisphere observers get some even greater objects, such as the brilliant globular clusters 47 Tucana (NGC 104) and ω Centauri (NGC 5139) and the colorful open cluster NGC 4755, commonly called the Jewel Box Cluster. Dropping down a notch on the impressiveness scale, the NGC and IC lists include easily overlooked treasures like the Owl Cluster NGC 457 in Cassiopeia, the Eskimo Nebula NGC 2392 in Gemini, the huge open cluster IC 4665 in Ophiuchus

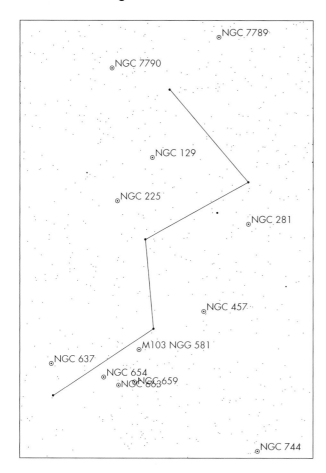

NGC 7789

NGC 7790

NGC 129

NGC 225

NGC 281

NGC 457

M103 NGC 581

NGC 637

NGC 654
NGC 663 NGC 659

NGC 744

Figure 3.4.
Lightweight programs such as *Stargazer's Delight* only show the brighter deep sky objects, as with this view of the constellation Cassiopeia; but if you observe with binoculars or a small telescope, you probably won't need much more detail than this anyway.

and the bright open cluster IC 2392 in Vela. The well-known British astronomer Sir Patrick Moore created a supplementary list to the Messier Catalog including some of these great NGC and IC objects. This list has become known as the Caldwell Catalog (Moore's surname is Caldwell-Moore, and since M1 to M110 are used for the Messier Objects, C1 to C100 make up this new list). While by no means an exhaustive sortie through the best of the NGC and IC lists, if a light-weight planetarium offers the Caldwell Catalog along with the Messier, so much the better!

Even amateurs with relatively modest observing equipment will appreciate a program that shows where the brightest satellites of Jupiter and Saturn are. A 76-mm (3-inch) refractor will easily show the four Galilean moons of Jupiter, Ganymede, Callisto, Europa and Io, and the Saturn's biggest moon, Titan. A 114-mm (4.5-inch) reflector will let you see a couple more of Saturn's moons, and a 200-mm (8-inch) SCT will show you six. However, seeing the moons isn't the same as knowing which is which, and for that you need a planetarium program that plots the position of these satellites around their planet. Some programs do this simply by drawing the moons onto the simulation, so that if you magnify the

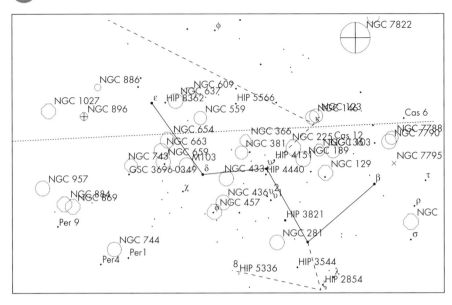

Figure 3.5. Heavyweight planetarium programs such as *XEphem* can show deep sky objects and stars beyond the thirtieth magnitude, allowing advanced amateurs to hunt down very obscure objects. On the other hand, all this detail can be confusing to newcomers, and unless you have very dark skies and a big telescope, such detailed star charts aren't much use to even the most experienced observers.

Figure 3.6. Many programs allow the user to zoom into the sky simulation sufficiently so that the positions of planetary satellites are apparent, as here with *Starry Night Pro.*

Figure 3.7. A few programs have separate windows for displaying the positions of planetary satellites, as here with *XEphem*.

region around the planet sufficiently, the moons become visible. Others have specific windows that show the moons around their planet.

Using a planetarium program as a substitute for a star chart or astronomical atlas can work nicely, although as we've already seen, moving your eyes between an illuminated laptop screen to the telescope eyepiece is a great way to mess up your dark adaptation. Many programs offer a night vision mode, which makes the screen much dimmer and with everything drawn in shades of red rather than the normal mix of red, green and blue. So configured, you can use both the laptop and the telescope comfortably and effectively. There are others things to look for as well. Especially useful are field of view indicators, usually configurable to match your eyepiece collection. These will help correlate what you see on the screen with what you can see as you look into the eyepiece. Many programs will

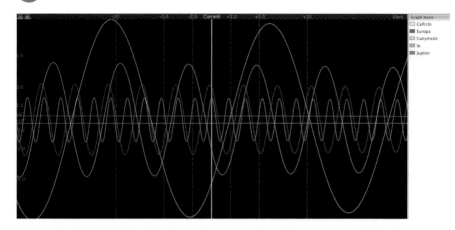

Figure 3.8. Although difficult to read, elongation charts for planetary satellites are very useful for predicting events such as eclipse of moons across their planet's disc and times when one moon will pass in front of another. Some planetarium programs can produce these charts, for example *Starry Night Pro.*

allow you to store a whole set of such indicators, one for each eyepiece in your toolbox. Since astronomical telescopes will show the image back to front, and Newtonians will show the image upside down as well, having a simulation that can be reversed and inverted can be very useful too, making the image on the screen much more like what you can see through the telescope.

One of the most attractive planetarium programs is Fabien Chéreau's *Stellarium* sky simulation (Windows, Mac and Linux; freeware). Although not an amateur astronomer's star charting program *per se*, amateurs using a small telescope or binoculars might find it a worthwhile map of the sky for locating bright stars, planets and so on. The program has some serious limitations though. For one thing it lacks a night vision mode, although as we've seen there are utilities that can implement this on a system-wide basis so this isn't really too much of a problem. More seriously, it has only a limited selection of deep sky objects (basically the best of the Messier list), doesn't plot the satellites of Saturn or Jupiter, and doesn't have field of view indicators (although the field of view of the entire screen is shown). On the plus side though, this is a beautiful application capable of producing stunning sky simulations. For an observer wanting to identify stars and constellations, and maybe find some of the showstopper deep sky objects for binoculars like the Orion Nebula or the Beehive in Cancer, this could be a perfectly usable program. Significant steps up in power are some cut-down versions of the commercial programs popular with more advanced observers: *TheSky Student Edition*, *SkyGazer* (based on Carina's *Voyager III*) and *Starry Night Beginner*. *TheSky Student Edition* and *Starry Night Beginner* are budget priced and available for both Windows and the Mac, while *SkyGazer* is (inexplicably to me at least) nearer the price of a heavyweight application and Mac-only to boot. All these applications look and feel a lot like their bigger brothers, just with a much more limited feature set. They are limited to the Messier Catalog of deep sky objects and star catalogs or around a hundred thousand stars. On the other hand, you can enhance any of these applications by combining them with an astronomy handbook including detailed constellation maps or descriptions of star hops to objects not on the Messier list (a few are listed in Appendix 1). Any of these programs will help the observer find the fairly bright stars used for star hopping to deep sky objects, since most star hops use stars down to about eight magnitude or so, and then the book can be used to finish the hunt via the finder-scope or low power eyepiece. In fact, such an approach mixes the best of both worlds, allowing the beginner to get help from the computer program for identifying specific stars in constellations (something much less easy than many experts would have you believe) while still providing scope for development of the star hopping skills. Even if you have a go-to telescope, learning the night sky is a great skill to have for when the batteries run down, when you're using another non-go-to telescope or binoculars, or simply want to show yourself or someone else something quickly without going through the rigmarole of aligning the telescope.

An alternative to a cut-down program is one specially designed for use with small telescopes and binoculars. Typical of these are Phil Harrington's *Touring the Universe Through Binoculars* (for Windows) and Ruedi Schmid's *Stargazer's Delight* (for the Mac) both of which are priced at the budget end of the market.

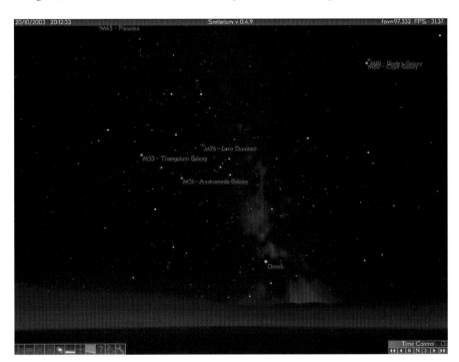

Figure 3.9. Photorealistic sky simulations are a hallmark of *Stellarium*, a basic open source program for Windows, Mac and Linux.

Though they have a similar feature set to the commercial programs already mentioned, they have a very different feel to them. *Touring the Universe Through Binoculars* is very definitely aimed towards observers who prefer to use binoculars, or for that matter small short focus refractors. Rather than just including the entire Messier list, it has a veritable potpourri of targets from all sorts of lists, each of which is a worthwhile observing target for astronomers using binoculars (or small aperture telescopes). Other goodies including field of view indicators and a night vision mode, things the three cut-down applications mentioned earlier lack. *Stargazer's Delight* is a planetarium program for families and classrooms. Whereas *Touring the Universe Through Binoculars* features tailored to a specific sort of observer, *Stargazer's Delight* comes with little presentations that explain various astronomical events such as eclipses. One nice extra to *Stargazer's Delight* is that unlike the programs mentioned so far, it includes a tool for plotting the positions of the moons of Jupiter, giving it a degree of utility for solar system observing the others lack.

Middleweight Planetarium Programs for the Backyard Astronomer

Middleweight planetarium programs are a big jump ahead of lightweight applications in terms of their capabilities: expect features like million-plus star databases, the full Messier and NGC/IC catalogs, night vision modes, and plots of the moons of Jupiter and Saturn as standard. Many will come with power-user features like go-to telescope control, user-expandable databases, Internet resource linking and customizable field of view indicators as well. All this of course comes at a price, and many of these programs cost two or three times as much as a commercial lightweight planetarium program, though a few are budget-priced shareware or even free. The other disadvantage is that these applications are much more complicated than the lightweight planetariums, and with a few notable exceptions none are truly easy to use, family-friendly programs. Bigger star catalogs mean busier sky simulations, making it more difficult to see constellations and asterisms, and under suburban skies especially, to decide which stars are actually visible for real and which only on the laptop display. Having said all that, a good middleweight planetarium program really can be all the software you're going to need. Relatively few amateurs need the truly vast star and deep sky catalogs sported by the heavyweights, not to mention the specialized tools like CCD integration and FITS image viewing.

RedShift (Windows and Mac) is probably the most limited of these middleweight programs and consequently not widely used by amateurs. Nevertheless, it is popular in schools and as an education resource for families, so is worth including. It does include a basic planetarium with the Messier Catalog of deep sky objects (the NGC/IC catalog is absent), plots of the Jovian and Saturnian moons, a map of the Moon and other such goodies. Even so, *RedShift* does betray its heritage as an educational rather than practical tool: it lacks a night vision mode, field of view indicators and go-to telescope control. For all that though *RedShift* is worth a look, if only because it can often be picked up at bargain prices in software retail stores and will amuse the children long after the clouds have rolled in and the telescope been packed away. François Nguyen's *Alpha Centaure* (for Windows) is an interesting freeware planetarium that is much more obviously a step above the lightweight programs though still relatively modest in scale and fairly easy to use, making it a much more attractive option than *RedShift*. For example, it has a rich roster of catalogs including many deep sky, asteroid and comet catalogs. It also offers Hertzsprung–Russell diagrams and planetary details like the position of the Great Red Spot on Jupiter, two features that are hallmarks of the best of the heavyweights. *Alpha Centaure* is suitable for use in the field, having a night vision mode, which is just as well, as notably absent from its feature set is a chart printing (although screenshots can of course be taken and pasted into something like *Paint Shop Pro* and printed from there). Somewhat similar is Microproject's *Equinox* (for the Mac), a low-cost, sophisticated shareware program that like *Alpha Centaure* offers all the basics as well as

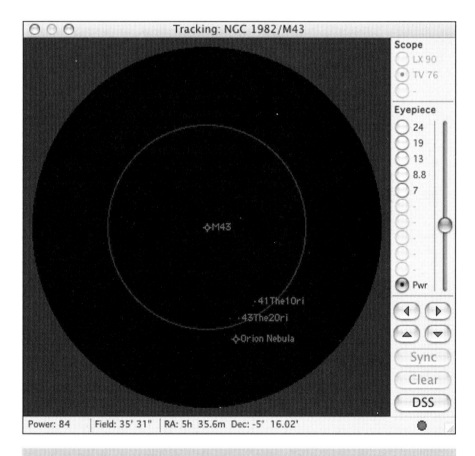

Figure 3.10. The Scope View window of the Macintosh planetarium program *Equinox* is one of its most compelling features, allowing the user to configure views to match a whole set of eyepieces and simulate their performance on three different telescopes.

some power-user features as well. In the case of *Equinox* the user can enjoy go-to telescope control, support for additional databases beyond the Messier and NGC lists it comes with, and observing lists that the user can create beforehand and use to guide the telescope once in the field. *Equinox* offers user configurable field of view indicators as well as a rather novel Scope View window that mimics what can be seen through the eyepiece at the same time as the main window shows the full star chart. This is very useful if you like to have your sky simulations looking as they do to the naked eye but the high-power views reversed and upside down to match the view through a telescope. Better still, this window works with a go-to telescope as well, and can be used to update the alignment and tracking of the telescope, by centring a target in the Scope View and then physically centring that target at the eyepiece.

For Linux users comes *KStars* from the KDE KStars team led by Jason Harris. Like most Linux software, this is open source, which for the end user means it's free. Even better, it also happens to be a very good program. *KStars* integrates well with the K Desktop Environment, connecting to various Internet resources through the *Konqueror* web browser, for example. Among other things, this allows *KStars* to download images from the Digitized Sky Survey at the click of a mouse. In common with UNIX and Linux applications in general, *KStars*

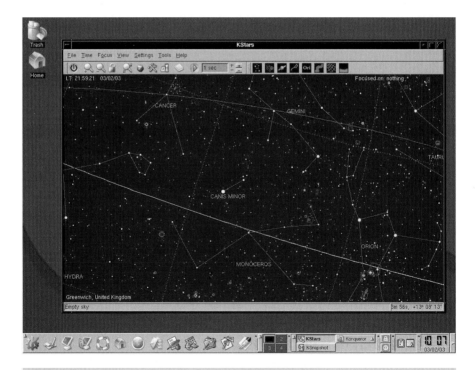

Figure 3.11. Although built for Linux, with an appropriate X Windows server *KStars* gives Mac users a perfectly serviceable, zero-cost planetarium program. Windows users will probably prefer the freeware program *Cartes du Ciel*, however.

produces PostScript printing files rather than bitmapped ones, resulting in noticeably sharp output when these files are printed using a laser printer. Though designed for Linux, *KStars* will run on a Mac (most easily installed using Fink) and on a Windows PC via an X Windows emulator such as *Cygwin*. There is more on installing and running X Windows software later in this chapter.

Starry Night Backyard is one of the most popular commercial middleweight programs for both Windows and Mac computers. Although it is underpowered as far as deep sky catalogs go (having only the Messier list), it does sport some advanced tools including plots of the moons of Mars, Jupiter, Saturn and Uranus. A simple yet powerful tool organizes the objects on view into tailor-made observing plans each night by adjusting various criteria such as magnitude, height above the horizon and object type. *Starry Night Backyard* is also one of the most attractive and realistic simulations, though this does come at substantial hardware requirements. The Mac-only program *Voyager III* lacks the planning facility and the eye candy, but is otherwise much more powerful, coming with many more databases including the full NGC/IC list, a night vision mode, and field of view indicators. It also has some neat extras like animated meteor showers and simulations of double stars for those who like following the progress of the secondary around its primary. One of my particular favorites is *The Digital Universe* (Mac and Windows) which shares a similar feature set to *Voyager III* but also comes with a superb astronomy encyclopaedia that is almost worth the price tag alone, not to mention a three-dimensional star simulation (complete with the necessary red and green lens spectacles!).

Heavyweight Planetarium Programs for Advanced Amateurs

Standard features to all these heavyweight programs are things like plots of satellites around all the major planets, field of view indicators and go-to telescope control. These applications also come with big deep sky databases, for those people for whom the NGC/IC, let along the Messier list, is just not enough. Many of these programs include deep sky object databases running into the hundreds of thousands, from galaxies and planetary nebulae through to quasars, not to mention thousands of asteroids and comets, and tens of millions of stars. If this whets your appetite, then read on!

Chris Marriott's *SkyMap Pro* (for Windows) has developed an enthusiastic following not least of all for its relatively modest price tag, huge complement of databases and advanced features, and the frequency with which the author updates and improves the program. Besides go-to telescope control, *SkyMap Pro* also works with digital setting circles, allowing the sky simulation to center on wherever the telescope is pointing. *SkyMap Pro* has very sophisticated observing session planning tools that allow the user to choose objects by class or magnitude for example, and an observing log for keeping a record of your achievements. Patrick Chevalley's *Cartes du Ciel* (for Windows) is also popular with many amateurs, and given the fact it does all the things a heavyweight planetarium needs to, but costs nothing to download and use, it is easy to understand why. One of the

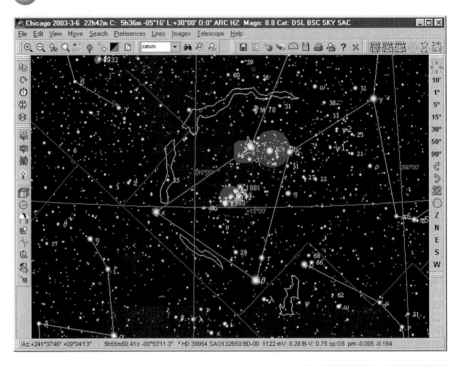

Figure 3.12. Despite being free and the work of just one programmer, *Cartes du Ciel* is a very powerful and easy to use planetarium program for Windows.

nice extras that this program sports is the ability download Digitized Sky Survey images from the Internet and then paste these onto the sky simulation, allowing you to make photo-realistic images like those seen in *Starry Night Pro*. *Cartes du Ciel* also has tools for accurately aligning photographic images by identifying stars in the image with those in the chart, useful with both DSS images and ones that you have taken yourself with a CCD or film camera. Integration with web resources is very good: you can update and expand the deep sky and star catalogs easily, get the latest coordinates for Jupiter's Great Red Spot, and download new asteroid and comet files.

Two very slick and widely used programs are *TheSky Level IV* and *Starry Night Pro* (both available for Windows and the Mac). Astronomers of any level of expertise can use and enjoy these programs, but where they have really established themselves is among astronomers of intermediate skill levels, offering a balance between ease of use and utility. Both are expensive though, and *Starry Night Pro* requires an especially powerful computer on which to run if it is to really do its stuff. *TheSky Level IV* includes big deep sky and solar system catalogs like many of the others at this level, but what sets it apart is its suite of integrated CCD control and image manipulation tools. Note that the Mac version lacks these tools (though given the paucity of Mac-compatible CCDs they are somewhat redundant), and is a good deal cheaper. In its latest incarnation, *Starry Night Pro* has become one of the most beautiful of all the planetarium programs, some would

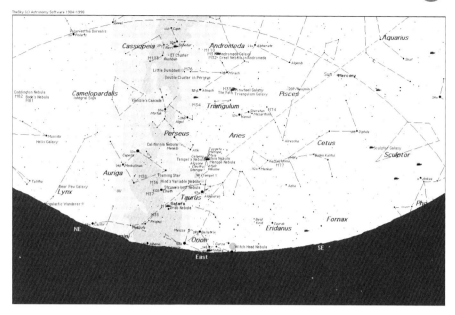

Figure 3.13. *TheSky* is something of a standard in the field of Windows and Mac planetarium programs, sporting a terrific array of tools, and is an ideal choice for intermediate and advanced amateur astronomer.

say it is has traded usefulness for entertainment value, but really that isn't true. While it lacks CCD and image manipulation software, it does have the same observation-planning tool as *Starry Night Backyard*, and links to Internet resources including the Digitized Sky Survey and Space.com. It also comes with an array of catalogs if not quite in the same league as *SkyMap Pro* or *MegaStar*, certainly for most observers enough to fill a lifetime of starry nights. *Starry Night Pro* also has something for lunar observers: move the cursor over features on the lunar surface and their names appear. Admittedly, this isn't a high-resolution mapping tool, but it is enough for casual observers to identify the seas and even relatively small craters. A distinct disadvantage is that this feature doesn't work in reverse, so you can't use it to locate features by name or control your telescope to point at them.

At the top end of the range are three highly respected commercial programs, *DeepSky, MegaStar and XEphem*. While even a casual observer would enjoy a heavyweight like *Cartes du Ciel* or *Starry Night Pro* (and many do), the sheer depth and complexity of applications like *MegaStar* can be baffling and intimidating. Surprisingly perhaps these monsters are not the most expensive of all the applications mentioned here, all three are less expensive than *Starry Night Pro*. Steven Tuma's *DeepSky 2003* (Windows) is a very complete package that pretty much covers all the bases as far as visual observers are concerned. It has colossal databases with over seven hundred thousand deep sky objects, arranged in a spreadsheet format making them very easy to browse while creating observing programs as well as use within the planetarium. This is a great feature if you are devising a very specific observing program, perhaps only one class of object, or all the objects within a certain area of the sky. There are various tools for planning observing sessions and recording your observations as well, not as simple notes, but in a powerful and flexible database format. Astronomers with CCDs or webcams will find the image enhancement tools useful. Overall, *DeepSky* is a remarkably rich suite of tools for the advanced amateur and considered by many of them to be one of the very best. Willman–Bell's *MegaStar* (Windows) has almost as large databases of deep sky objects and like all the programs at the top end of the heavyweight range, it is relatively easy to find and add new databases. Like *DeepSky*, *MegaStar* plots large deep sky objects not as symbols but as shapes much more similar to what you will see, so an open cluster will be made up of stars, and a nebula will have distinct, shaped outline. Many other programs, even high-end applications like *Starry Night Pro*, tend either to use symbols or photographs, which although pretty are not always accurately positioned, and so can be misleading. Again, *MegaStar* has earned itself a very devoted following among advanced observers, particularly those who observe challenging objects like faint galaxies and nebulae. Unhappily, for Mac and Linux users, *DeepSky* and *MegaStar* are both Windows-only applications (although they run fine in Windows emulations software). Mac and Linux users do have a high-end program of their own, Clear Sky Institute's *XEphem*. It is similar to *DeepSky* and *MegaStar* in charting abilities except that plots of deep sky objects are symbols and not drawings. Very distinctive features of *XEphem* are the windows it uses to show the surfaces of Mars and the Moon. These are photographs of these worlds

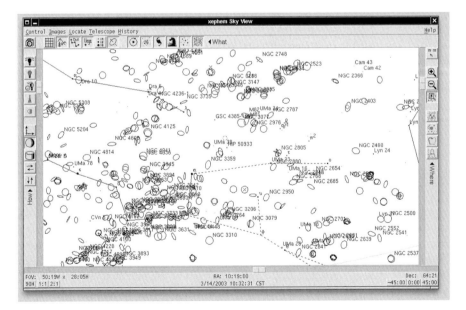

Figure 3.14. *XEphem* is typical of the top-end programs that have truly huge databases of objects including hundreds of thousands of deep sky objects: you'll run out of clear nights long before you see them all!

showing things like the terminator, geological features and spacecraft landing sites. As with UNIX programs generally, *XEphem* produces PostScript files for printing and inclusion in publications, making it an ideal choice for astronomers needing to put star charts into newsletters, magazines or books where bitmapped images simply won't do. Like *KStars*, *XEphem* works most easily with UNIX operating systems such as Linux and the Mac, but with a suitable X Windows server, it will run on a Windows PC as well, though most PC users prefer applications like *TheSky*, *MegaStar* and *DeepSky*.

Using Planetarium Programs

Once you've chosen which planetarium program is best for you, it's time to look at what it can do. To begin with, it is always fun just snooping about, reading the instruction manual and generally putting a program through its paces. Many programs are fun as well as useful, but here the accent must be on the latter criterion: how can a planetarium program help you observe more objects more easily? Some people like to use their laptops out in the field beside the telescope (with night vision mode engaged of course). For beginners especially, having an interactive star atlas that changes as the evening wears on is a great boon. You can easily find stars by typing their names into the search box in the program, or identify the moons of Saturn and Jupiter by zooming into those planets as if you were hitching a lift on *Voyager II*. Some observers find that even with the laptop display set to its minimum, the light it casts impairs their night vision sufficiently to make the difference between seeing a faint deep sky object or not at all. In this case, it makes much more sense to print off your star charts ahead of time and use these instead. Others just don't like bringing an expensive, fragile and power-hungry piece of equipment into the field. This is especially an issue if you observing location is somewhere miles from the nearest city, like a national park or desert. The best skies are often in such remote locations, where charging a battery isn't going to be easy, and inclement weather or accidental damage can quickly turn a laptop into high-priced paperweight.

Producing and Using Star Charts

The alternative then is to use your planetarium program as the source of your charts rather than the chart itself. Virtually all planetarium programs have a sky-charting white sky mode replacing the realistic colors of the planetarium with a much more diagrammatic interpretation of black stars on a white background. In fact, many programs will print off the chart in black and white even if the screen simulation is in full color. Of course, you'll still need some sort of light to see them by, but a weak astronomy flashlight (often utilizing low-power red light-emitting diodes) casts much less light than laptop display, and is much less detrimental to your dark adaptation.

Why produce and print off your own charts at all when there are many star atlas books and maps already published? One reason is that with a planetarium program you can print off a whole series of maps suited to each part of your observing program. The analogy is with geographical maps: some maps will cover the whole world, another the continental Unites States, a third the state of Nebraska, and a fourth downtown Lincoln. Theoretically, you could find Lincoln on the map of the world, but it wouldn't help you find your way around the University campus. Conversely, if you wanted to see how far Lincoln, Nebraska was from Lincoln, England, the downtown map of Lincoln wouldn't be any help at all. As with maps, you need a star chart of the right scale, or expansiveness, for the task. Identifying the constellation of Leo from its neighbors requires one sort of star chart, covering a large area but without so many stars plotted such that the figure of Leo becomes lost (making it impossible to see the constellation for the stars!). On the other hand, star hopping from Denebola to M61 will need a much finer resolution map but one covering a far smaller patch of sky. What seems like plenty of stars in the large region covering Leo won't be enough at this higher resolution, and the faint stars that were ignored before will be needed to match the view through the finderscope or eyepiece to make the star hop successful. Even with a small telescope you'll find yourself using seventh and eight magnitude stars routinely to get at many deep sky objects, particularly in regions relatively devoid of bright stars (the Great Square of Pegasus being one of the best known such regions).

Now consider your astronomy books. You might have one of the pocket sized ones that include stars down to fifth or sixth magnitude with small maps seemingly crammed with stars. These are fine for identifying the constellations, but at the eyepiece or even a decent finderscope, you will see many more stars than these show you. Therefore, they aren't very helpful for star hopping. Full size star atlases offer more space and clarity, and usually more stars as well since they include stars fainter than eighth magnitude or so. The other side to this is that these atlases are more expensive and somewhat baffling to beginners, containing a vast amount of information that can make it difficult to identify the basic constellations if you aren't sure what you're looking for. In fact many star hoppers like to use two or more atlases, one for getting to the approximate location from some other distant point in the sky, and then a different, higher resolution atlas to use at the eyepiece or finderscope while zeroing in on the target. Catalogs and other books include descriptions of targets arranged by constellation or type, and though these may be inspirational they rarely include maps detailed enough for star hopping. A planetarium program can reduce the need for weighing down your bookshelves by giving you a way to find objects to observe and then produce the successively more detailed maps covering smaller and smaller areas of the sky you need to find them.

So how is the best way to do this? Most programs will automatically toggle between a high density of stars at high magnifications and a low density of stars at low magnifications. The key thing is that at each magnification you have a sufficient density of stars to correlate the map with what you can see, while at the same time not having too many stars such that the useful asterisms of stars makes the star hop difficult to follow. For naked eye observations, for example when you are aligning your telescope towards a bright star conveniently located near to the deep sky target being sought, stars down to third or fourth magnitude are probably will probably be fine. These are the sorts of stars that make up the constellation patterns, and should be visible even under moderately light polluted suburban skies (sadly, in the worst city skies sometimes not even the very brightest stars can be seen). Figure 3.15 shows two star charts produced by *TheSky*, successively zooming into a region of the Milky Way between the bright constellations Cygnus and Aquila containing a couple of rather small but interesting constellations, Sagitta and Vulpecula. This region of the sky contains many curious and fascinating things, including many wonderful star clusters and nebulae for astronomers using small telescopes. With increasing magnification, the amount of space covered by the chart gets smaller, but more and fainter stars and deep sky objects show up.

Let us put this into practice. Our quarry will be the Coathanger Cluster, a striking asterism of forty or so stars that really does look like a coathanger (some books also refer to this cluster as Brocchi's Cluster or Collinder 399). This cluster is also quite easy to find, and so makes a great example for taking a look at what sort of charts are needed to find deep sky objects and other small, faint astronomical objects. The obvious place to start is at the bright star Altair, an easily distinguishable star thanks to two slightly dimmer stars, Alshain and Tarazed that flank Altair and make a nice tight row easily seen with the naked eye. Beginning at Altair (near the left-hand edge of Figure 3.15), moving the two degrees to Tarazed is an easy enough star-hop. The next bit is the seven degrees north towards two bright stars at the end of the Sagitta "arrow", α and β Sagittae. Because these two bright stars are close together they are easy to spot, and they lie about halfway between Tarazed and Albireo (β Cygni). The final step is from these two stars to

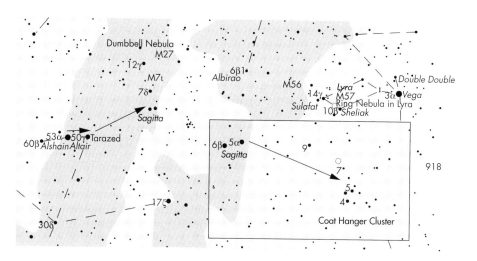

Figure 3.15. With each successive increase in magnification, a smaller portion of the sky is shown, but with more details on view. In this example, star charts at two different magnifications help create a star hop from Altair to the Coat Hanger Cluster.

the Coathanger Cluster itself, which lies about four degrees northwest of β Sagittae (shown in the small inset in Figure 3.15).

Admittedly, this is a very easy star hop but the technique is universal and widely used by observers the world over. Before we leave star hopping as a concept, there are two other great things about planetarium programs that make them so very useful for this sort of thing. Firstly, many planetarium programs allow the user to invert and reverse the sky simulation turning it into what the observer sees through the finderscope or eyepiece. This isn't needed for the initial naked eye view, or for a low power view using a pair of binoculars, because you'll see everything the right way around and right side up anyway. But if you are using an astronomical refractor or catadioptric telescope what you'll see will be a mirror image of reality, while a Newtonian reflecting telescope doesn't just reverse the image, it turns it upside down as well. Most planetarium programs offer a reverse and inverting facility that allows you to manipulate the star charts so that they match the eyepiece image more closely. The second great feature sported by many planetarium programs is the eyepiece field of view indicator. These circles mark off the region of the sky visible through a given telescope eyepiece, making it much easier to reconcile what you see on the screen with what you are looking at through the telescope. Depending on the program used, these field of view indicators are either fixed in the center of the display and you move the simulation about "underneath" them, or else they are dropped onto the simulation and stick to it, allowing you to line up a series of them mimicking an entire star hop step by step. This latter type is particularly useful if you want to print off a single chart but with a series of eyepiece views shown, rather than use the laptop alongside the telescope in the field. The trick with these sorts of charts is to overlap each eyepiece view with a distinctive star or asterism, so that as you move the telescope along you don't lose your bearings. So if your first view has an obvious double star on the northern edge, then that would be a good place to position the southern edge of the next eyepiece view.

Devising Observation Programs

An observing program is simply an itinerary of astronomical targets for the observer to carry out along the course of an evening. Many amateurs never bother, preferring to view things serendipitously, simply seeing what constellations and planets are in view, and choosing targets from among them. For solar system objects like the Moon, Jupiter and Mars whose appearance changes quickly, this approach is fine because there is always something new on view. On the Moon, the terminator is of most interest, and as its position changes continuously, so every night promises a different view and new features to explore. Jupiter and Mars also change rapidly, if a bit more subtly than the Moon. With Jupiter, it is the movement of the four Galilean satellites and the banding on the planetary disc, particularly the Great Red Spot, which holds the eye. When watching Jupiter, it is almost possible to feel that you looking at it from a spaceship (admittedly rather far off) watching the planets and its moons spin and whirl beneath you. That feeling is even more intense when observing Mars, at least during those precious few weeks every couple of years when it is so close to the Earth that surface features, ice caps and dust storms can all be seen even with a small telescope.

Deep sky objects are different. They don't change at all on the human timescale, and if you just look at a handful of the same objects night after night, even showstoppers like the Great Orion Nebula, you'll eventually get bored with them. Some go-to telescopes have built-in tours of the "night's best" objects that the onboard computer calculates on the fly, and though a bit limited in variety (they tend to concentrate on solar system and Messier objects) these can be one way to expand your observing repertoire. Many magazines and books give observing lists tailor made for certain times of the year, but with many planetarium programs you can go even further and create your own lists. These automated planning tools work by calculating which objects will be visible that night and what time they will be seen at their best, and then apply various filters to remove some of the objects from the list according to criteria the user sets up beforehand. Sometimes these filters serve a very practical purpose; for example, if you live where houses, trees or mountains obscure the horizon you might want to remove objects less than 30° above the horizon. You might also want to use filters to remove objects fainter than a certain magnitude because ambient light pollution would prevent you seeing them anyway. Another use for filters is to narrow down the range of objects to your sky conditions. You might choose to concen-

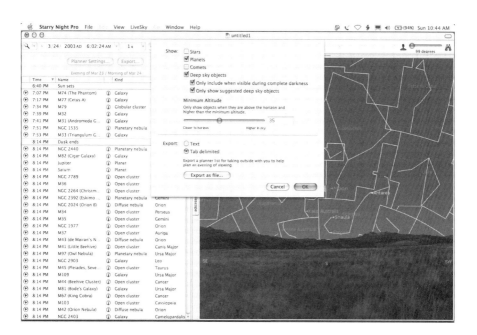

Figure 3.16. *Starry Night Pro* includes a simply but effective tool for devising observing sessions that make best use of your time. Export these as text files for printing if you don't want to take your computer into the field with you.

trate on deep sky objects rather than stars or planets on dark night, whereas on a night with a full Moon it might be better to stick to double stars and planets. Filters help focus your observing session onto a certain class of object, for example galaxies or asteroids. You might want to do this for purely practical reasons: a small aperture, short focal length refractor is a great tool for observing open star clusters and large emission nebulae, but a bad choice for most galaxies and globular clusters. Conversely, a large aperture SCT will deliver excellent views of globular clusters and galaxies where aperture and medium to high magnification are the order of the day. Nevertheless, you may also want to concentrate on a certain type of solar system, star or deep sky object that interests you. Indeed, even if you don't, an evening spent looking at just galaxies or globular clusters for example can be very worthwhile because it helps you recognize and understand the different classes of these objects and their structure.

The planning tool in *Starry Night Pro* is typical of those built into many planetarium programs. Itineraries can be filtered and sorted, and then either used inside the application or printed off for use with traditional star charts and books. For example, you might want to sort the objects by type and declination, concentrating on galaxies no less than 40° above the horizon to maximize your chances of seeing them. Applying fine filters gives you much more control over the observing program than the relatively coarse filters included in the main simulation window itself. Once you have made your observing program, you can print it off (together with some star charts) and you're all set for the evening. Although there are many books that include lists of deep sky objects, arranged by constellation (as is the case with *Burnham's Celestial Handbook* for example) or the time of year when they are best observed (as with *Turn Left at Orion*). My particular favorite though is the *Field Guide to the Deep Sky Objects* by Mike Inglis. This book classifies objects first by their type, the best time of the year to see them, and finally by how easy or difficult they are to observe. Using a book like this it is easy to put together a short observing program for each night. For example, you might decide to look at a four-day old Moon, Saturn and then Mars one evening because they are well positioned in the sky. Once the Moon has set and you have had your fill of the planets, then you can work through a short observing program of globular clusters, getting to know these fascinating objects better. Half a dozen might be fine for a night, but should give you ample opportunity to look at the differences in density and compactness, the ease or otherwise of their resolution, and their general shape. Incorporating this sort of semi-serious program into your serendipitous stargazing is a great way to keep your hobby fresh and exciting.

Using Planetarium Programs to Locate Objects by Day Using Setting Circles

The setting circles of equatorial mounts, as mentioned in Chapter 1, sound a simple way to "dial up" deep sky objects but in practice many amateurs never use them. The budget mounts in particular have rather crude setting circles and are difficult to accurately align and calibrate to begin with, and even if you understand the theory, actually using the setting circles on these mounts can be frustrating. Because these mounts are often provide newcomers to the hobby with their first experience of setting circles, many astronomers wind up having tried using them a couple of times, given up, and never going back to them even when they can afford much better quality mounts with more accurate setting circles. But setting circles can be fun, and even with very basic mounts like the CG (or EQ) family of mounts that come with the entry-level telescopes from makers like Celestron and Orion, you can at the very least play around with them and try your hand at this alternative to star hopping. Always fun to do and surprising to those who get a look through the telescope is using the setting circles to find planets in broad daylight. Besides the cleverness of being able to find something you can't see with the naked eye, using the setting circles by day allows you to see and understand what you're doing. This makes this "daytime astronomy" a great way to try out setting circles before using them by night to hunt deep sky objects or whatever else interests you.

Understanding Mounts and Setting Circles

The typical equatorial mount that comes with amateur telescopes, particularly refractors and Newtonian reflectors, is the GEM or German equatorial mount. The mount has two axes of rotation, RA, or right ascension, and declination. The right ascension axis is the one with the polar alignment telescope threaded through it (if you're mount has this accessory, the smallest and most basic mounts don't). Around the right ascension axis is the right ascension setting circle, and it will be marked off in units of time. The crudest setting circles are marked off in relatively course divisions, usually only the hours are numbered and tick marks between each hour for ten minute intervals (this is the case with the mounts that come with the Celestron Firstscope series, for example). More sophisticated setting circles have finer divisions and a Vernier scale (as is the case with the popular Vixen mounts) to allow you to make readings down to one-tenth of a minute. The more accurately you can read setting circles, the easier they are to use. The declination axis is the one with a counterweight at one end and the telescope, within some sort of adjustable cradle, at the other. The mount sits on the tripod, around which the mount revolves in azimuth.

When correctly set up for observers in the northern hemisphere the right ascension axis should point to the northern celestial pole, around which the skies appear to rotate. This point corresponds almost exactly to the position of the bright star Polaris, making the job of alignment in the northern hemisphere quite straightforward. In the southern hemisphere there are no bright stars so close to southern celestial pole, only a relatively dim star, Sigma Octanis, so aligning the right ascension axis towards the southern celestial pole is a bit more difficult. Either way, once set up the right ascension axis should describe an angle to the ground equal to your latitude. It is possible for the latitude scale to be incorrect. Check this using a spirit level (the best sort of mount have these built in, for a few dollars you can finds small ones designed to fit into the eyepiece holder of the telescope – these are especially handy). The "0" value should correspond with the horizontal and the "90" with the vertical. You may need to change the lengths of the tripod legs to get the mount level if you are on sloping or uneven ground.

Properly set up, any celestial object of known right ascension and declination coordinates can be located by simply rotating the telescope around both axes to the desired right ascension and declination coordinates. You can get the right ascension and declination coordinates of an object from various sources including books, astronomy magazines and planetarium programs. Normally the telescope rotates around this right ascension axis, and then the declination axis, northwards for positive values and southwards for negative values. If your telescope has a motor on the right ascension axis, then that axis will turn in time with the object in the eyepiece, keeping it in the field of view (if accurately polar aligned, adjustments in declination are usually not necessary for visual observing, at least not for observations of a few tens of minutes).

Daytime Use of Setting Circles

One of the great fun exercises for anyone using a telescope on an equatorial mount is to hunt down astronomical objects by day. Of course, this requires proper polar alignment; but how do you do this if you can see the stars? One way is to set the telescope up at night and come back to it during the day, but another way is to use the Sun. *Be warned that the Sun is a dangerous object to look at directly* – through a telescope the heat and light can be magnified so much that they can cause irreparable harm to the eye. Therefore a proper solar filter must be used, that is, one that has been designed and sold for use for visual observation through astronomical telescopes. There are various sorts produced for photographic cameras, binoculars or naked eye observing – these are not appropriate for telescope use.

To begin with, simply guess which direction is north or use a compass to roughly align the telescope, and then set the declination to match your latitude. The next step is to find out when the Sun will be due south. A planetarium program will show you this if you run the simulation and see when the Sun passes over the southern cardinal point (which is usually plotted on the horizon or as part of a grid system over the sky simulation). This is the time known as local noon, and before the standardized timekeeping across nations necessary with the building of the railways, this was the fixed time point around which local people arranged their days. At Greenwich, England, local noon occurs at about 12:00 GMT while at Bristol about two and a half degrees (about a hundred miles) west it is ten minutes later. At this time, shadows will be pointing due north, and you can use these to align the telescope northwards more accurately than with a compass, which of course points towards the *magnetic* and not *celestial* north pole. This is also how a sundial works: the shadow cast by the gnomon points north at local noon. In the southern hemisphere, the same principle applies but the shadows point due south instead.

Now comes the fun bit. Making sure the solar filter is fitted securely, turn the telescope towards the Sun. Notice that the shadow the telescope casts is as small as it can be, and incidentally, once you've seen this it becomes quite easy to point the telescope towards the Sun for making solar observations without using a finderscope. Use the planetarium program to establish the right ascension and declination coordinates of the Sun. Assuming you have the tripod set on a flat, horizontal surface the declination axis shouldn't need much adjustment, but it is quite probable that the right ascension coordinate is significantly off. This just means your polar alignment of the telescope using shadows was a bit off. Usually revolving the mount in azimuth slightly around the tripod should make this correction. Turn the mount a degree or so at a time, re-center the Sun in the field of view and take another look at the right ascension setting circle value, and repeat until it has the right value. Basically you are using a known set of right ascension and declination values that you can "measure" by observation, those of the Sun, to fine-tune a second set of right ascension and declination values, those of the northern or southern celestial pole, that you can only estimate indirectly using a compass and shadows. With luck you should now have a mount with the right ascension axis pointing due north or south as required, even though you can't actually see the guide stars you would use to do this by night!

With the telescope so calibrated, you can now take up the challenges of looking for objects seemingly invisible by day. The Moon is a great first target because you may be able to see it with the naked eye as well. Find its right ascension and declination coordinates, dial them up with the setting circles and then look through the finderscope or a wide-angle eyepiece. Obviously, don't forget to remove the solar filter before looking, but take great care if you're looking at the Moon less than three days from new not to get the Sun in the field of view by accident. How young (or old) a Moon can you find?

Though not visible to the naked eye, even the finderscope should be enough to let you see Venus and Jupiter by day if the telescope is pointing in the right direction. Again, use your planetarium program to find their coordinates. What you can see depends on the aperture of the telescope, but for a small telescope like a 114-mm (4.5-inch) reflector, nice views of Jupiter are perfectly possible, with at least a couple of the bands being apparent. A bigger telescope such as a 200-mm (8-inch) SCT should allow you to see Mercury and Saturn as well. Mercury in particular is worth hunting for because when observed by day you can see it through much less atmosphere than is the case when it is observed at dusk or dawn close to the horizon (a necessity for dark sky observations of this little planet). In fact this is what the greatest visual observer of Mercury, Eugène Antoniadi, did while mapping its surface during the 1920s.

Other Useful Computer Programs

So far, I've concentrated on planetarium programs and the sorts of things they do. Although by far the most significant portion of the software market aimed at amateur astronomers, there are many other programs that amateurs are likely to find useful or at least interesting. Many of the programs describe in the following section is shareware or freeware, and links to the download sites can be found in Appendix 1.

Moon-Mapping Software

After a good planetarium program, many amateurs find a computerized lunar atlas are one of the most useful additions to their toolkit because they do something that books don't do that well. The Moon changes its appearance continuously, and between the effects of its phase and libration (the way it seems to wobble from side to side), the illumination of craters and other features is incredibly variable. A crater at one extreme of libration can look completely different to how it looks at the other extreme and in between. Books are usually limited to a few plates or diagrams, and sometimes it is difficult to match what you see at the eyepiece with the diagram – so how are you supposed to know what you're looking at? *Virtual Moon Atlas* (for Windows) is a freeware program from Patrick Chevalley, author of the *Cartes du Ciel* planetarium program. It superimposes the names of features onto a correctly illuminated map of the Moon, making identifying lunar features much more simple. This program doesn't produce maps suitable for printing because they are bitmapped, screen-resolution images, but on the screen of a laptop, they would work very well. A more sophisticated commercial program called *Lunar Map Pro* (Windows) does produce vector-image,

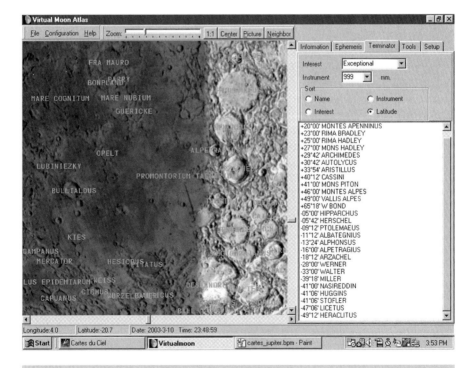

Figure 3.17. *Virtual Moon Atlas* helps lunar observers find features including craters and seas and includes various tools for filtering searches. This high-magnification view shows the terminator approaching the Mare Nubium.

high-resolution printed maps, and so would be a better choice for someone looking for a source of maps to print off and use at the telescope that way. Either of these programs would make a great supplement to a planetarium program though, most of which lack a proper map of the Moon.

List-Based Planning Software

Although we've looked at planning software as incorporated into planetarium software, list-based planning software has become quite popular too, especially for use with go-to telescopes. Like planetarium programs, these programs can drive a go-to telescope automating the movement from one object to the next, but these programs eschew the graphic point and click approach favoured by planetarium programs and instead of finding objects on a sky simulation the user puts together a list and works through that. At their best, these sorts of programs are much faster to use that planetarium programs, and more importantly they give the user a place to store observing data, making them an ideal choice for those who want to record their observations and thoughts on the things they look at. They also record some sorts of data automatically, such as the time of observation, which can be great if you're doing something like a Messier marathon, and want to keep a log of your evening's work. Ilanga's *AstroPlanner* (Windows and Mac) program is a powerful, multi-purpose planning tool. It allows the user to maintain a database of objects and observations, identify the objects in the eyepiece field of view, adopt a red-screen night vision mode, correct for polar drift and calculate the optimum pairs of stars to use for go-to telescope alignment. Depending on the computer and telescope it can do some remarkable things, such as allow the user to control the telescope by voice, plot the temperature of the optical tube and thereby establish when it has cooled down sufficiently for high-resolution imaging. Stephen Hutson's *Scope Driver* (for Windows and Mac) is another of these list-based observing log programs with a similar sort of feature set, but is perhaps a bit leaner and easier to use. In many ways, it feels like an expansion of the go-to telescope handset, with more bells and whistles, but the same sort of experience. If you enjoy using a go-to telescope for skipping between deep sky and solar system objects quickly but want to enhance your tours of the night sky as well as record notes as you go along, either of these tools could easily replace a planetarium program.

Astrophotography Software

One class of application worth mentioning here are graphics programs designed especially for use by amateur astrophotographers such as *AstroStack* (Windows) and *Keith's Image Stacker* (Macintosh). Chapter 6 covers their use in detail. As the names of the two examples mentioned suggests, among their various functions these programs stack images, in this case frames from a webcam movie to produce a single high-resolution image. Essentially what they do is extract the frames, carefully align them on top of each so that the edges of the object being

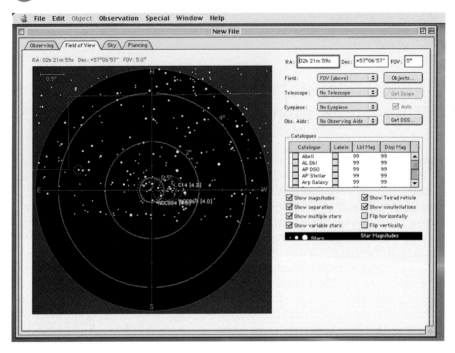

Figure 3.18. The simulated eyepiece view of *AstroPlanner* gives the user an impression of what should be visible through the telescope complete with labels. One nice touch is the availability of different eyepiece settings, allowing you to switch magnifications and fields of view depending on the eyepiece.

imaged lines up precisely, and then squashes the whole stack together so that details vaguely caught in each frame are enhanced and much more obvious. As well as this function, they also offer tools for improving images still further by using sharpening and (perhaps surprisingly) blurring algorithms of various sorts. Although not a complete substitute for graphic editing software like *Adobe Photoshop* (Windows and Mac) these program do automate and simplify the particular things that astrophotographers need to do to enhance their images. Where the more general-purpose graphics programs like *Photoshop* really come into their own is for creating composite pictures made up from a series of smaller ones. This is usually necessary with the Moon, for example, where only a small portion of the lunar disc is contained in any one frame of the webcam movie. Stitching together numerous frames each containing a small region of rather like a patchwork quilt lets allows you to create a much bigger image, called a mosaic, than a webcam could otherwise record in one shot. Being able to tweak each frame so that it matches the brightness and colors of the adjacent ones is essential if the joins between frames are to be invisible. Incidentally, many of the well-known pictures of the planets taken by space probes like *Mariner* and *Voyager* are mosaics as well, the original images covering only small strips or squares of their targets and requiring much the same handling to be turned into the bigger portrait shots.

Professional, Academic and UNIX Software

The Internet is a good source of programs designed for teaching astronomy or for carrying out scientific research. Although most of these programs have limited value to amateur astronomers as far as the practical aspect of their hobby is concerned, many of them add greatly to the educational side of the equation. Typically, such programs are freely downloadable, but there are some downsides. For one thing, the bulk of academic and research-grade software is source code for use with UNIX workstation computers. Linux and advanced Mac users will have the fewest problems with this as they will be familiar with the three key steps of "compiling", "making" and "installing" applications downloaded as source code, but Windows and casual Mac users may find the whole process rather baffling. Fortunately, some of the most popular programs in this category come as ready-made installers for those computers (known as binaries) that generally streamline the operation significantly. This is the case with *XEphem*, *DS9* and *FITS Viewer*, for example.

The problem with binaries is that they are usually specific to a certain variety of UNIX, for example Linux running on an Intel processor or X11 on the PowerPC. Source code is, in theory at least, usable by any UNIX machine regardless of hardware or operating system, but instead of the binary offering a quick installation, the computer must compile the code from scratch, which takes much longer. Life is rarely that simple though, and altering source code to get it to compile on a certain type of hardware, a process called porting, is horribly complicated. Luckily for neophytes to UNIX programming, the applications described in this section are available either as binaries or as source code ported to the most popular computer types. Fink is one such project, where Mac users can download ports of popular Linux and other UNIX applications (see Appendix 1), but there are numerous others.

Installing and Using X Windows

The other major hurdle to jump is the requirement for an X Windows server, as for example, required by *XEphem* and *KStars*. X Windows is a graphical front-end to a UNIX computer, and provides the basic interface of the Linux desktop and application environment as well as a secondary interface to the Mac operating system. Don't confuse *X Windows* with *Microsoft Windows*; these are very different things. A PC, for example one running Windows XP, isn't using a UNIX-based operating system (as do Linux or Mac computers) but one that is proprietary to Microsoft and based on their early MS-DOS operating system. As such, a PC can't run UNIX or X Windows software *natively*, that is, at full speed, in the same way as Linux or Macintosh computers. Instead, X Windows software needs to run in emulation, that is, with X Windows running on a *make-believe UNIX computer* created by an *emulation program*. Though remarkably effective, software running in emulation is always slower than software running natively because the emulator must translate instructions between the software and hardware instead of letting them communicate directly as with native software. The other problem is that sometime the emulation is imperfect, particularly with regard to peripherals like printers and USB devices, which might not work with the emulated software at all.

In use X Windows looks a lot like the Microsoft Windows user interface, with programs running in windows and each one having menu bars. There are also things like task bars and keyboard shortcuts, and in practice getting used to X Windows doesn't take very long. As mentioned before, Linux users will already be using an X Windows interface, and so will already be running X Windows programs (the screenshot of *KStars* earlier on in this chapter featured an X Windows server on a Linux computer). Mac users also have a UNIX machine, but their usual front end, Aqua, isn't an X Windows server. The simplest solution for them is to download and install a program called *X11* available on Apple's web site. This creates an X Windows server that runs alongside the normal interface and within which X Windows programs run. There are alternatives, including using *XFree86* with or without *OroborOSX* (these two packages come with the CD-ROM installation of *XEphem*, a great boon to Mac users without an adequate Internet connection to download the big files required).

Life is a little more complicated for Microsoft Window users. They need to install both the X Windows software plus a program that can emulate the UNIX computer needed to run it, such as *Cygwin, WinaXe or Virtual PC*. The practical details of installing these fall outside the scope of a book like this but links for these programs given in Appendix 1, and these sites that have plenty of information on the process and what's required. Once in place though, you can use X Windows programs almost as easily as can Linux and Mac users. The other alternative for Windows users is to install a second hard drive (or partition their existing one) and install a complete Linux package such as from Red Hat or SUSE. That way, the user can choose to start up the computer running either Microsoft Windows or Linux (with X Windows built-in) as needs be. Because Linux works well on low-specification machines compared to Windows XP or any of the other current consumer-grade operating systems, creating a Linux machine is also a great way to breathe new life into an old desktop computer that you wouldn't otherwise have much use for. Though the ready to go Linux packages are commercial products (and consequently come with instruction manuals and CD-ROMs filled with all the files you need), it is possible to download an entire Linux operating system from the Internet, if you have a fast enough connection. A 56k modem simply isn't up to the job, but with something like DSL or cable, you can expect to download the files in around an hour. Even so, if you do fancy going down this avenue, spending a little money on a Linux handbook of some sort will be a sound investment.

GeoVirgil – Exploring the Solar System from Your Desktop

Amusing and informative as planetarium programs like *Starry Night* are, they don't reflect the full depth of our knowledge of the planets, particularly those in the inner solar system that have been imaged in detail by a whole series of probes. *GeoVirgil* is step towards giving interested amateur astronomers a taster of what NASA and other space scientists have access to, a frame-by-frame view of the surfaces of Mars, Venus, Mercury and the Moon.

GeoVirgil was developed by Steve McDonald and is based on an open source mapping toolkit called *OpenMap* used primarily by scientists that being based on the "virtual computer" system known as Java can run on any computer able to run Java programs. Most modern desktop operating systems can do this, so downloading and running a program like *GeoVirgil* is easy. Although early Java programs had a reputation for being slow and a bit flaky at times, this is no longer the case and *GeoVirgil* should run on your computer perfectly well.

The program is easy to use. From the File menu the user can summon up images by choosing to input a name, or coordinates, or simply browse a type of feature on a given planetary body. Once selected, the program accesses the files

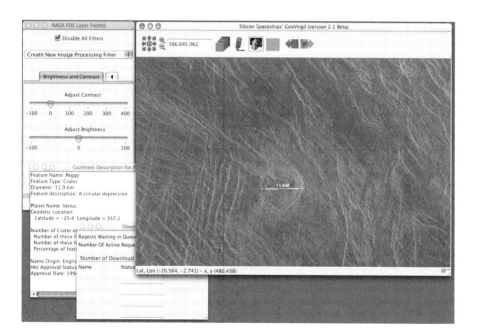

Figure 3.19. *GeoVirgil* is a Java-based program that will run on any modern computer with a fast Internet connection. It downloads and displays images of the surfaces of the planets together with key information about the thing being viewed and performs a variety of additional functions such as allow the user to measure the dimension of structures and export the picture as a JPEG.

across the Internet, loads them into its memory, and displays it in the main window. There is usually a bit of information about the object or region in one of the additional windows alongside the main image. Filters tweak the image brightness and contrast, which can be useful with those features that don't show up clearly at the default settings. If you want, you can save the file or export it as a JPEG. Extra buttons allow the user to zoom in or out, label features, move across to neighboring regions and download views of them, and so on.

FITS Viewer and DS9 – Viewing *Seriously* Deep Sky Images

Professional astronomers lean heavily on a file format known as a Flexible Image Transport System (or FITS) file that is able to combine image data alongside all sorts of extra numerical and statistical data. These are the files they share with one another taken with professional telescopes. Needless to say, the quality of these images is astounding and far better than anything most home astronomers will ever hope to achieve, if only because the professionals use really big telescopes that are able to image very faint objects. Unfortunately, most graphics packages are not able to view these files, which is a shame as many of them are publicly accessible (if you know where to look) via the Internet. Fortunately, there are a variety of special programs that allow the user to locate and look at these images, one of which is called *fv*, or *FITS Viewer*, is free to download and install, and runs in an X Windows environment with an Internet connection.

One of the nicest things about *FITS Viewer* is that it comes with options for accessing files across the Internet without the user actually needing to know the web sites or FTP sites from which the FITS files will come. For example, you can search the Digitized Sky Survey with criteria such as NGC number or the galactic coordinates of the region of interest. *FITS Viewer* will then download and display the image part of the file, and if you want to look at the rest of the file contents, other windows will show those as well. An alternative to *FITS Viewer* is *DS9*, Again, this is a professional astronomer's program designed for use with X Windows, and many of the same functions and features. However, it is more for image analysis than file viewing, and has some useful educational resources built-in. One of the best of these is the Virtual Observatory that can connect to a variety of servers that offer informative texts and guides that go alongside the images that *DS9* downloads and displays. Another nifty feature is the way *DS9* can access the Digitized Sky Survey to download images (using an option under the Analysis menu). Basically all you do is type in the name of the object (for example NGC457 or Omega Centauri) or the coordinates of the region of space you're interested in. It is important to keep the width and height parameters sensible: make them too small and the object you're interested will be occupy the whole window and it'll look like you've downloaded white fuzz. So if you're trying to get a picture of something big like the Orion Nebula, and it doesn't seem like *DS9* is displaying anything, check you've set the width and height parameters right. My particular favorite feature is being able to browse an archive of data server across the Internet using the delightfully named "non-astronomers interface" that takes you

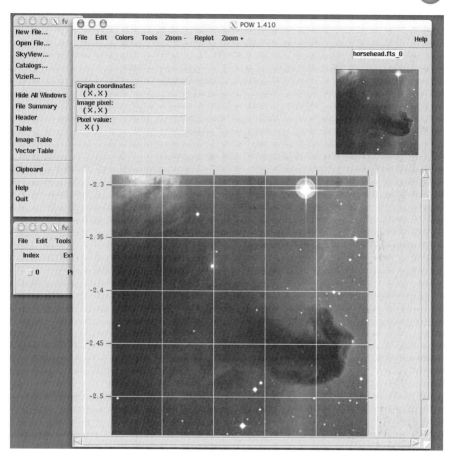

Figure 3.20. *FITS Viewer* opens and displays the files professional astronomers use to exchange image data, and though hardly essential to the amateur, it is sometimes fun to see what something like the Horsehead Nebula looks like through a big telescope.

straight to the pretty pictures and shows you what terms to enter to summon up the original images into *DS9* (or *FITS Viewer*, for that matter). Both *DS9* and *FITS Viewer* can export the FITS files as files that other programs can open. Do this using an interpreter program such as *GhostScript* (installed by default in some X Windows installations) or by simply printing to file, which means the image is saved as a generic PostScript file that a graphics program like *Photoshop* or *The GIMP* can easily open and convert to a JPEG or bitmap.

Nightfall – Modelling Binary Stars

Nightfall is for those astronomers with an interest in binary stars, particularly those that are eclipsing variables as well. Once again, *Nightfall* requires an X

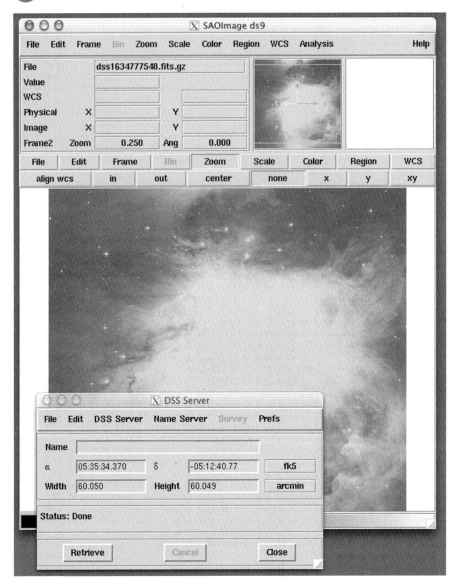

Figure 3.21. *DS9* is an alternative to *FITS Viewer* that can access online repositories of deep sky images and educational resources.

Windows environment to run in, and if you want the interactive animations (and you do!) then a UNIX graphing program like *Gnuplot* or *PGPlot* is required. Although somewhat intimidating to configure, install and use, author Rainer Wichmann includes plenty of explanatory notes along with the source code which should allow most users to install *Nightfall* without too much trouble. There is also a Mac version at the Fink project.

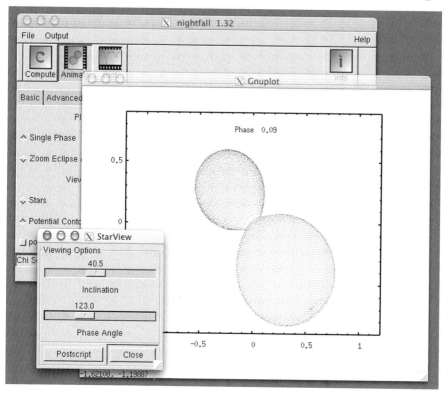

Figure 3.22. A simulation of how the close binary system DD Monocerotis might appear produced by the X Windows program *Nightfall* by Rainer Wichmann.

The graphical display is based around an animated view of the two stars orbiting one another, alongside of which is plotted the changes in magnitude, or lightcurve (in the screenshot give here, it's the graph at bottom right). Changing factors such as the size and temperature of the stars or the observer's angle of view is where most casual users will find *Nightfall* the most fun. If the two stars are identical in size and temperature, then when they pass in front of each other there will be two identical dips in magnitude. However, if one star is of different size or temperature, then the changes in magnitude will be different. This is exactly what happens with the famous variable star Algol in Perseus. *Nightfall* also calculates any distortions in the shape of the stars: some stars are so close to one another than instead of being spherical they become teardrop-shaped, as the hot gas is attracted from one star to another, and in some cases the stars may even touch. *Nightfall* includes a feast of other output options and data plots, enough to keep the serious double star enthusiast amused for hours.

StarPlot – Mapping Our Corner of the Universe

Another X Windows program, *StarPlot* by Kevin McCarty offers an unusual way of looking at the stars in our region of the galaxy, plotting them in three dimensions around a given star (such as our Sun). Changing the magnitude limit or the diameter of the simulation in light years makes more or less complex charts, as does changing the number of stars included. Useful additional features added such as the names of the stars and "stalks" showing the position of the star above or below the galactic plane. A nice addition is a graph showing the magnitude of the stars included in the chart against their spectral class, the well-known Hertzprung–Russell diagram. *StarPlot* doesn't show constellations or deep sky objects, and so is in no way a substitute for a planetarium program, but it is an amusing toy for the amateur interested in the three dimensional distribution of the stars around our home planet. It's also a useful educational tool, showing information on the nearby stars such as spectral type and distance from Earth.

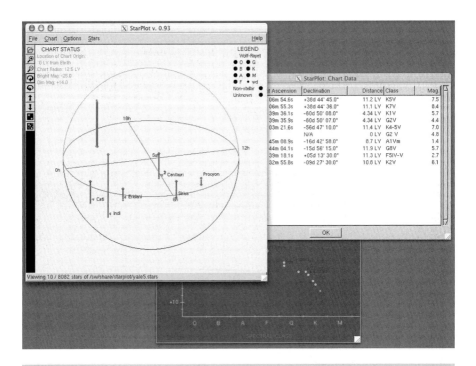

Figure 3.23. Kevin McCarty's *StarPlot* program is an easy-to-use stellar cartography tool for exploring the distribution of stars in our region of space.

SETI@Home – Contributing to the Search for Alien Life

We can see the light waves travelling through space using our telescopes, and radio telescopes allow professional astronomers to do the same thing with radio waves. Many astronomical objects produce radio waves, including things as diverse as fierce storms inside the gas giants, solar flares and the super-massive black holes in the hearts of certain types of galaxy. Radio waves also come from artificial sources here on Earth, obviously radio and television transmitters, but also things like computers, cellphones, and other electronic devices. No other animals on Earth make radio waves. If we could find an artificial-looking radio wave coming from somewhere in outer space, that would be compelling evidence of intelligent life. The trick is telling an artificial radio wave from a natural one. In Carl Sagan's popular science fiction book *Contact*, the aliens used a series of prime numbers to get our attention, and this is indeed one of the things scientists studying radio waves in the search for alien life look out for. A mathematically significant series of numbers would be unlikely to be a by-product of some natural process and so would imply intelligence behind the signal. SETI, or the Search for Extra Terrestrial Intelligence, is the name given to the largest and most systematic survey of different radio frequencies for artificial signals that could be our first indications of alien civilizations. The problem is that radio telescopes are

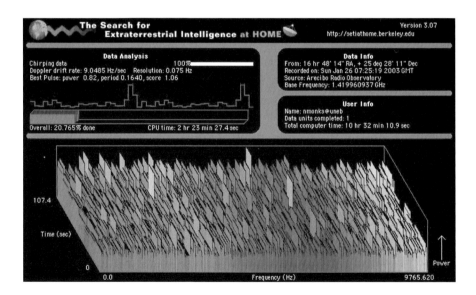

Figure 3.24. If you leave your home computer running when you're at work or at night, then maybe *SETI@Home* is one way for you to help professional astronomers do serious work. The program looks and acts like a screensaver, but is really searching for mathematical patterns in data collected by radio telescopes that could be a good clue that there is intelligent life in space.

capable of gathering a vast amount of radio wave data relatively quickly, and the scientists working on the project simply don't have the computing power to sift through all the data to pick out the artificial signals from all the background noise.

This is where *SETI@Home* comes into play. This project uses computers in homes and offices to process SETI data, feeding any observations of radio signals that might be worth looking into more closely back to the scientists at the project headquarters. Versions of the *SETI@Home* program work on Windows, Mac or Linux computers, the only requirement being a connection to the Internet. This connection doesn't need to be permanent, the program can download a packet of data when you're online and then work through it happily when you're offline, but it will of course need to be reconnected to send the results back to the main SETI computers and to download the next packet of data. Under the guise of a screensaver it springs into life whenever the computer the computer isn't being used for anything else, for example at night, and can even be run as a background application while you're doing word processing or surfing the Web, if you have a fast computer.

CHAPTER FOUR

Buying a Go-To Telescope

Go-to telescopes come in one of two varieties: those designed and built as all-in-one robotic telescopes, and those that began as traditional manual or motorized telescopes but had computer control added afterwards. There are advantages and disadvantages to both types. The all-in-one telescopes are remarkably easy to set up and use, but on the other hand, the range of optical tubes is limited. Most of the middle and high-end models are SCTs or Maksutovs, for example, rather than apochromatic refractors or larger aperture Newtonians many advanced hobbyists prefer. The go-to upgrades produced by Losmandy, Vixen, and others are relatively expensive accessories and rather more difficult to assemble. Their key advantage is that they are much more flexible, and a single upgraded mount will work as well with a small refractor as a large Newtonian, assuming that it is mechanically up to the job. Moreover, since these upgrades work with the mount and not the optical tube, you can change the optical tube as required. Despite this, the all-in-one models dominate the market. They make it possible for even the most inexperienced amateurs to see lots of solar system and deep sky objects during a night's observing. Compared to setting circles or star hopping the skills needed are minimal and are quickly: find north or south, identify a few bright stars and then use the computer handset to align the telescope. All-in-one go-to telescopes have been particularly popular with newcomers to the hobby looking to buy their first telescope, for which the two major manufacturers, Celestron and Meade have provided models ranging from cheap and cheerful 60-mm short-focus achromatic refractors through to respectably sized 200-mm (8-inch) SCTs capable of doing serious observing. For the higher end of the market, both Celestron and Meade offer large aperture all-in-one SCT telescopes up to 400 mm (16 inches) in size as well as a smattering of other designs.

For all their popularity with beginners, the wider amateur astronomy community has been somewhat snobbish about go-to telescopes. Old timers maintain that since the views through the eyepiece can never match those of photographs of deep sky objects, newcomers will be disappointed if they use their go-to telescopes expecting to see a succession of brilliant astronomical spectacles. Instead, the entertainment value of the hobby comes not from seeing things through the telescope, but in developing the skills necessary to find them in the first place. These might include learning to read astronomical maps, recognizing the constellations and their principal stars, star hopping between stars and deep sky objects, and using the setting circles on a telescope's mount. Rely on just the thrill of seeing things at the eyepiece without adding on these technical skills and practical challenges, and the level of involvement and understanding needed to maintain a long-term interest is lost. Go-to therefore takes away from the user the opportunity to build up skills and put them to the test of finding things, essentially removing the challenge and so cheapening the rewards. It is all too easy for an observer using a small go-to telescope to find themselves running out of new and interesting things to look at (and let's face it, one faint smudge looks much like another, whatever fanciful name it might bear). Once that happens, it isn't long before the telescope finds itself packed away in the garage or in the basement, and the would-be amateur astronomer moving on to other things.

This chapter, and the one that follows it, are about how to prevent that happening. I believe that go-to telescopes offer casual newcomers to the hobby every opportunity for success with the minimum of fuss, and that has to be a good thing. Many of the people who buy go-to telescopes (or even get them as presents) are people who wouldn't otherwise get into the hobby, or if they did get started, would give up soon after. The small go-to catadioptric telescopes especially are neat and very portable, attractive in a high-tech sort of way, and very easy to use. Assuming you can at least identify the alignment stars properly (and these are usually obvious first magnitude or brighter ones), then even a newcomer can expect to be bagging interesting deep sky and solar system objects from the word go. This is absolutely key to getting people to stay in the hobby, and to encourage them to expand their horizons from simply sightseeing through to really observing things: following the monthly observing columns in the astronomy magazines, taking photographs, recording what they see, and so on. Whether or not the experienced astronomers approve of them, go-to telescopes are here to stay.

Do You Need a Go-To Telescope?

This is a tricky question to answer, not least of all because until a few years ago practically everyone in the hobby got by fine without go-to on their telescopes, so clearly isn't an essential feature. Nevertheless, for many newcomers to the hobby the idea of a computer doing the difficult parts of the process of hunting down deep sky objects and automatically tracking the planets and the Moon is very attractive indeed. Whatever the merits of learning the sky over using a computer, there is a second issue to consider, and that is the extra expense. All else being equal, telescopes become more expensive the larger they are, but the larger a telescope the more objects it will show and the more detail you will see at the eyepiece. This is the "aperture wins" rule of buying a telescope: A top of the line 90-mm (3.5-inch) apochromatic refractor will be beaten hands down by a well-made but otherwise ordinary 150-mm (6-inch) Newtonian costing a tenth as much when it comes to brightness and resolution. Consequently, many experienced hobbyists recommend beginners start with a Newtonian of some sort, since this design more than any other offers the best compromise between size, cost and performance.

This produces a quandary at the bottom end of the go-to telescope market. While the optical tube, mount and tripod get less expensive as they get smaller, the computer and motors are the same regardless of the size of the telescope. Go-to might add just a few percent to the cost of a 400-mm (16-inch) pier-mounted SCT, but fifty percent or more onto the cost of an entry level refractor or reflector. On top of this it has to be remembered that although go-to makes it easier for the user to place an image of a star or galaxy at the eyepiece it doesn't make the image any brighter or more detailed: a 90-mm go-to telescope is still a 90-mm telescope as far the photons and your eyes are concerned. Moreover, a small telescope only delivers rewarding images of only a relatively small number of objects, principally the Moon and planets, compared to a 200-mm (8-inch) or larger instrument that will give nice views of galaxies, globular clusters, and more. Finding the sorts of objects a small go-to telescope will do well on isn't too difficult as most are obvious to the naked eye (no one needs a computer to find the Moon!). In contrast, the thousands of objects suitable for observing through a big go-to telescope can be hard work to find and go-to will help enormously. Therefore, while there is a good *practical* argument for adding go-to capabilities to a 200-mm (8 inch) aperture telescope, with a 90-mm (3.5 inch) one what you are paying for is *convenience*. The buyer needs to balance the extra convenience of having a small go-to telescope that finds things automatically with the greater versatility of a larger manual telescope that may be more difficult to use but will show more objects, more impressively.

Celestron or Meade? Autostar or NexStar?

This is a questions amateur astronomers debate endlessly. Certainly, there are some optical tubes in each of the manufacturer's line-ups that have been highly lauded over the years for their consistently high quality and optical performance, the Celestron 125-mm (5-inch) and 235-mm (9.25-inch) Schmidt–Cassegrains and the Meade 178-mm (7-inch) Maksutov being cases in point. Getting a go-to system built around any of these optical tubes would therefore be a particularly good choice. On the other hand, the computer technology behind the Autostar and NexStar systems is mature and reliable, and both systems are very popular among amateurs. The bottom line is to decide on the optical tube and mounting system that best suits your needs, and then find a model from either manufacturer that satisfies these criteria and is within your budget. Don't worry about whether it is an Autostar or a NexStar telescope. There are trivial differences between the two computer systems, certainly; for example, the software in the Autostar handset can be updated via a home computer and the Internet, but this cannot be done with the NexStar, but otherwise they both work in much the same way. If you want an all-in-one go-to telescope, either of the two designs will fit the bill very well.

All-in-One Go-To Telescopes

Go-To Short Focal Length Refractors

These are among the smallest of all the go-to telescopes and are generally inexpensive even before the discounts many stores offer for these introductory models. As with all short focal length achromatic refractors, these are useful instruments in their way, but they do have some serious limitations. For a start, these are telescopes for use in the same sort of way as binoculars. They are great

Figure 4.1. As with all the small achromatic refractors, the Meade ETX 70 performs indifferently where high magnifications are required, as with lunar or planetary observing. Consequently, it makes a poor choice for an absolute beginner looking for a multi-purpose instrument; where it excels as a second telescope for deep sky observing on trips to places where the Milky Way is bright and clear (photo courtesy of Meade Instruments Corporation).

for viewing things like big, bright open clusters and the expanses of stars that make up the Milky Way, targets on which traditional long focal length telescopes provided too narrow a view to show well. The other major limitation to these telescopes, at least in the achromatic form present in the go-to telescopes sold by Celestron and Meade, is that they perform badly at high magnification on bright objects. Chromatic aberration becomes apparent, resulting in spurious color around objects like the Moon, the planets and even the brighter stars. The degree to which this annoys the observer varies, but it is something to be aware of before going ahead and buying a high-magnification eyepiece or a Barlow lens to use on such an instrument. At best, chromatic aberration is merely annoying, but at worst, it can be very distracting.

Both Meade and Celestron have go-to telescopes of the short focal length type. The Meade version is a 70-mm (2.8-inch) f/5 instrument, the ETX 70, and it comes with a cut-down version of the Autostar computer seen in its larger go-to telescopes. It comes with a lightweight field tripod adequate for the visual observing at low powers that this telescope is suited for, but a finderscope is not included. This may frustrate some users, in which case a stick-on zero-power finder, such as a Telrad, would be an ideal enhancement. Meade expects most users simply to treat the entire telescope as a giant finderscope by using a low-power eyepiece (a 32-mm Plössl would be ideal). The optical tube is unlike the traditional achromatic refractors sold by Meade and its sibling companies. Quite

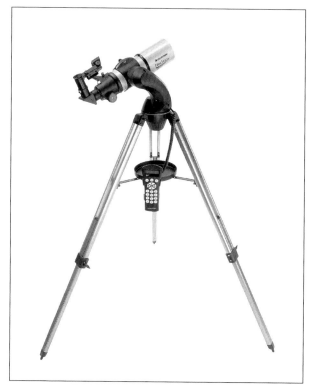

Figure 4.2. The Celestron NexStar 80 is a larger and more conventional-looking instrument than the Meade ETX 70, but otherwise shares many of the same strengths and weaknesses (photo courtesy of Celestron).

Table 4.1. Summary of short-focus achromat refractor features

Pros:	Refractors don't need to be collimated, and are compact and durable, making them ideal for travelling with. Small tubes cool down quickly, so are ready for use at a moment's notice. Lightweight tripod and mount performs well for visual use. Wide fields good for observing star fields and certain types of deep sky objects.
Cons:	Short focal length and achromatic optics not as good for high-magnification views of the Moon or planets compared with other designs. Small aperture means these telescopes are limited to views of the brighter deep sky objects, such as open clusters and large nebulae.
Ideal for:	Observers who routinely view from dark skies where the Milky Way and its constituent nebulae, star fields and open clusters are most obvious. Suburban astronomers looking for a small telescope to take on vacations to dark sky sites.
Not ideal for:	Astronomers wanting views of solar system objects free of false colour or wanting to look at deep sky objects while using light pollution filters, where an aperture of 150 mm (6 inches) or more is required.

different is Celestron's competing model, the Celestron NexStar 80, which uses the same Chinese-made wide-field 80-mm (3.1-inch) f/5 tube employed by many other telescope manufacturers including Orion and Helios for their traditional, non-go-to "rich field" telescopes. For all its limitations, this optical tube has been popular for many years, and it works well as a wide-field instrument. Again, this isn't a telescope to push too hard on the planets, at high magnification, chromatic aberration is significant, but 80 mm (3.1 inches) is a good aperture for taking in the detail on star clusters. To help with the initial alignment, the NexStar 80 does come with a simple but perfectly useful red-dot reflex finder. It also comes with a lightweight aluminum tripod. Despite the low cost and compact design of these two telescopes, neither is an ideal primary telescope for a newcomer to the hobby. Rather, they are something to use *alongside* a larger telescope such as an SCT or Newtonian, and perhaps doubling-up as a holiday telescope to be packed and taken along on vacations to places where the Milky Way is bright and clear and just waiting to be explored!

Long Focal Length Refractors

Meade produce a range of traditional achromatic telescopes on Autostar-equipped go-to equatorial mounts as part of the LXD 55 series (along with some Schmidt–Newtonians as well). Having a longer focal length that the ETX 70 wide-field refractor, chromatic aberrations are less intrusive and higher magnifications are possible without the need for very short focal length eyepieces. These telescopes are still a little on the fast side (f/8–f/9) to expect the best possible performance from the achromatic designs and there is still some unwanted color on bright objects, but most users find this a fair trade-off given the relatively low price of these instruments. The LXD 55 refractors include models with apertures

Figure 4.3. The LXD 55 series of refractors offer sufficient aperture to turn in decent views of deep sky objects under dark skies, but really come into their own as instruments for observing the Moon, planets and doubles stars. Some false color is evident on bright objects, however, and the lightweight tripods may need some modification, or even upgrading (photo courtesy of Meade Instruments Corporation).

from 125 to 150 mm, priced around $700 to $900, a positive bargain compared with the thousands of dollars it would cost to buy an apochromatic refractor of similar size and then upgrade a traditional mount to go-to operation. However, large refractors are unwieldy and heavy instruments, and the more demanding visual observer or astrophotographer will likely find the low-cost aluminum mounts and tripods Meade supplies these telescopes with a bit inadequate. Expect jiggles when focusing, not to mention the transmission of any vibrations from the ground up to the telescope. One solution is to use the anti-vibration pads sold by several of the telescope manufacturers. While these don't completely neutralize unwanted wobbles, they do attenuate them and reduce the length of time they persist for, making the telescope much more pleasant to use.

The jumbo go-to achromat in the Celestron product line is the C6 R-GT ($1000). This is essentially their 150-mm (6-inch) f/8 refractor mounted on a version of the CG5 equatorial mount that has been around for years, but now with NexStar go-to capability added. The optical tube has proved to be versatile and popular, and like the LXD 55, it represents an attractive compromise between optical performance and price. Although fine for visual observing at low to moderate magnifications, you may find the CG5 mount not quite stable enough for long-exposure astrophotography and high-magnification visual observing. Celestron also offers the $250 NexStar 60, a 60-mm (2.7-inch) f/12 achromatic refractor otherwise rather similar to the NexStar 80 mentioned earlier. Though lightweight and stable, this is rather too small for anything other than the most casual astronomical use. It's relatively long focal length doesn't make it particularly suitable for the sorts of deep sky work where its aperture might be adequate, such as Milky Way star fields and open clusters, and it's limited aperture won't

Figure 4.4.
Celestron's C6 R go-to telescope combines their popular 150-mm achromatic refractor with a computerized version of the CG5 equatorial mount, resulting in a versatile and reasonably stable instrument (photo courtesy of Celestron).

Table 4.2. Summary of long focal length achromat refractor features

Pros:	Collimation unnecessary and cool-down times relatively short. Focal length long enough to reduce chromatic aberration and allow high magnifications without requiring focal length eyepieces. Inexpensive compared with apochromatic refractors.
Cons:	Chromatic aberrations are still present and sometimes intrusive. 150 mm (6-inch) aperture is still quite modest compared with similarly priced Newtonians, too small for effective use of light pollution filters, particularly narrow-band ones, limiting their use as deep sky instruments. The optical tubes are big and heavy, and sometimes overwhelm the lightweight aluminum mounts provided.
Ideal for:	Beginner and intermediate level observers wanting a reasonably priced, general-purpose instrument. Lunar and planetary views can be good if you can ignore the false colour, and under dark skies, the quality of the image mitigates the limited aperture somewhat.
Not ideal for:	Deep sky enthusiasts needing a large aperture for galaxies, globular clusters, etc. Solar system observers requiring colour-free, high-magnification views that would be better suited by apochromats or Maksutov telescopes.

show much on the planets either. Really, this is rather more a toy than anything else.

Newtonian Reflectors

Reflectors have not been as popular among prospective purchasers of small and midsized go-to telescopes as the catadioptrics. The bulk of the average Newtonian optical tube compared with the much more compact catadioptrics and refractors probably has a lot to do with this. Celestron do offer a rather neat 114-mm (4.5-inch) short focal length reflector on the same single arm go-to mount as the NexStar refractors, called the NexStar 114 ($400). The aperture of this telescope is just enough to show some detail on deep sky objects like globular clusters, galaxies and planetary nebulae. As with all fast Newtonian telescopes, there is the possibility of coma, an optical aberration that turns stars away from the center of the field of view into v-shaped smears. A corrector lens just ahead of the eyepiece

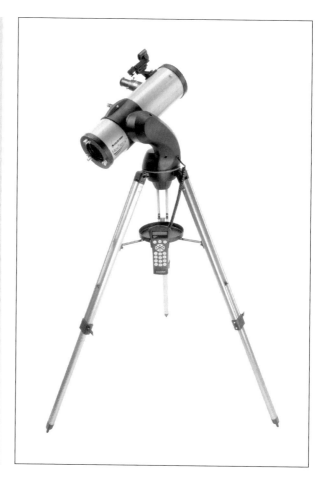

Figure 4.5. The NexStar 114 is a short-tube Newtonian similar to those used on a range of entry-level telescopes. Although modest, the aperture size on this telescope is just large enough to start being useful on deep sky objects as well as the Moon and planets (photo courtesy of Celestron).

mitigates this somewhat, but still, images aren't *quite* as sharp as those through a traditional long focal length Newtonian of the same aperture. Nonetheless, the NexStar 114 is an inexpensive, versatile and compact go-to telescope, and a better bet than the NexStar 60 or 80 as a beginner's instrument.

A healthy step up in performance (and cost) are a series of equatorially mounted Newtonians with the Autostar handset and motors bolted onto them, dubbed the Celestron Advanced series of telescopes. Equatorial mounts offer some important advantages over alt-azimuthal ones. Astrophotographers appreciate the fact the field of view doesn't rotate in the eyepiece as it does when telescopes track alt-azimuthally. Provided the tripod and mount are appropriate, visual observers can expect a stable image even at high magnifications, thanks to the counterweight. However, equatorial mounts aren't anything like as compact, and the Celestron Advanced series telescopes are all seriously big telescopes with heavy counterweights. The 200-mm (8-inch) f/5 model costs $900 and the

Figure 4.6. A good Newtonian is difficult to beat, and the combination of equatorial mountings and decent optics makes the Advanced Series reflectors from Celestron an attractive choice for the amateur wanting a versatile telescope at a reasonable cost (photo courtesy of Celestron).

250-mm (10-inch) f/4.7 about $1200, less than half the price of a SCT of similar aperture. Unlike the similarly priced C6R-GT refractor, these telescopes are ideal for deep sky work, combing a generous aperture with a wide field of view.

Comparable to the Celestron Advanced series is Meade's LXD 55 series. These are 150 to 250-mm (6 to 10-inch) Schmidt–Newtonians priced from $700 to $1000. While also quite fast telescopes (f/4–f/5), these telescopes have a correcting lens at the front that reduces aberrations at the edge of the field of view. Even better, these telescopes have focusers that can accommodate 2-inch (50.8-mm) eyepieces. Overall, these are mediocre performers as far as visual observing goes, what really sells them is their low price. Money saved on buying one of these telescopes over a SCT might go towards getting a couple of premium wide-angle eyepieces that will offer flatter, sharper views than generic Plössls or Kellners. Highly corrected eyepieces like Panoptics will mitigate somewhat a distinct disadvantage of these telescopes, coma, the tendency for stars towards the edge of the field to be v-shaped rather than true pinpoints. The short focal length can be a problem too; although great for observing large deep sky objects and star fields, it is not desirable when your target is the Moon or planets, where high magnifications are preferable. For example, to get a magnification of ×200 when using the 200-mm model, an eyepiece with a focal length of 4 mm is necessary. Even substituting this for an 8-mm eyepiece and a Barlow lens doesn't help much: Plössls and Kellners of this focal length have very tight eye relief and are generally unpleasant to use. The large central obstruction softens the images as well; expect SCT-like performance rather than that of a traditional medium or long focal length Newtonian. One final issue is the mounting that these telescopes come with. While adequate for the 150-mm model, the 200 and 250-mm models seriously burden the mount, exacerbating vibrations and making focusing at high powers difficult. Chapter 5 includes tips for improving lightweight aluminum mounts, but it is as well to realize from the start that these telescopes are not well suited to long-exposure astrophotography.

Go-To Maksutovs and the Smaller SCTs

The little 90-mm (3.5-inch) ETX 90 Maksutov from Meade is the telescope that spawned the whole go-to craze, and remains one of the most popular go-to telescopes available today. Unquestionably the compact design is a prime factor here: it can be used without a tripod, although it is much better on its special table top tripod or mounted on a sturdy camera tripod, and the whole thing easily fits inside a backpack making it one of the most portable astronomical telescopes around. With a price tag of about $600, it isn't cheap (this sum of money would get you a very much larger Dobsonian telescope, for example), but clearly many users consider that this is a reasonable price for such a compact, portable go-to telescope. Like all Maksutov telescopes, the ETX 90 is a long focal length instrument; so at f/15 this telescope is the complete opposite of the wide-field refractors mentioned earlier: this is a telescope that offers high magnifications and a narrow field of view. Fortunately, Maksutovs in general appear to be relatively easy to make consistently well, and even at high magnification users can expect crisp,

Figure 4.7. The LXD 55 series includes a range of Schmidt–Newtonian telescopes. These do offer plenty of aperture at a reasonable price, but they do have some shortcomings: so-so visual performance and less than rock solid mountings (photo courtesy of Meade Instruments Corporation).

contrasty images of the Moon and planets, and despite the large central obstruction that could impair resolution, these telescopes are suprisingly good at resolving double stars. Resolution and contrast is excellent, and at their best, these telescopes perform almost as well as a top-notch refractor. Couple the go-to with automatic tracking, and you have a great telescope to take out onto the porch to

Table 4.3. Summary of Newtonian reflector features

Pros:	No false colour. Relatively inexpensive compared to refractors or cata-dioptrics of the same aperture. Long focal length Newtonians can deliver high-magnification images of the Moon and planets of similar resolution and contrast to those of refractors, and better than SCTs. Medium and large aperture models work well with light pollution filters.
Cons:	Slow to cool down compared with refractors, though much faster than catadioptrics. Large equatorial mountings need to be robust and heavy for an acceptable degree of stability. Short focal length Newtonians only deliver good high-magnification views if well made, otherwise aberrations such as coma become intrusive. Only collimated telescopes deliver optimal performance.
Ideal for:	Any observer will enjoy a well-made Newtonian telescope. Long focal length telescopes are the best for planetary and lunar observing, and thus would suit suburban astronomers especially well. Short focal length ones better suited to low power, wide-field observations under dark skies. In either case, per unit of aperture Newtonians offer the best value and thus performance per dollar spent.
Not ideal for:	Observers requiring lightweight, portable and zero maintenance instruments; they will find other designs more convenient, though more expensive.

search for craters on the Moon, watch the satellites of Jupiter, or hunt down pretty double stars like Albireo. As with the ETX refractors, the pointing accuracy of the Autostar is good when set up properly and on a stable tripod, but on a lightweight tripod, it is all too easy bump these telescopes out of alignment, and the narrow field of view exacerbates any errors. Furthermore, the 90-mm aperture is still a bit too small for the observing galaxies and globular clusters, while the narrow field of view is less suited to large open star clusters than the wider field of the ETX refractors. You can expect nice views of the Orion Nebula and similar large, bright objects, but it is very easy to have the Autostar computer turn the telescope towards a deep object that is essentially invisible through the eyepiece (this can easily give the impression the ETX 90 isn't working). Another problem with the ETX 90 is the built-in finder, widely considered to one of the worst ever made, having a uselessly small aperture (24 mm) coupled with far too high a magnification for its size (eight-power). Mounted awkwardly as well, a zero-power finder like a Telrad would be a tremendous boon to most users, and make aligning the telescope much easier. Scaled-up versions of the ETX 90 are also available, the 105-mm (4.1-inch) ETX 105 and the 125-mm (5-inch) ETX 125. These have more comfortable, though still rather small, right-angled finders and of course more aperture to work with, making them more versatile as well. These two scopes cost around $900 and $1100 respectively, including a reasonably sturdy field tripod (a heavy duty, alt-azimuth and equatorial mount tripod is available for $300 or so, an essential purchase for astrophotographers). A big step above the ETX series is the sole Maksutov design among the otherwise all-SCT LX 200 range. This 178-mm (7-inch) instrument resembles a typical 200-mm (8-inch) SCT from the LX 200 series as described below, but with its even longer focal

Figure 4.8. Although the ETX 90 Maksutov does have its problems, its optical performance is excellent and there's no doubt it is a small and highly portable telescope ideal for quick looks at the Moon or to take on trips abroad (photo courtesy of Meade Instruments Corporation).

length, f/15 compared to the f/10 of the SCTs, it is especially suited to visual observing and imaging of relatively small objects like planets. As with the ETX Maksutovs, these telescopes can deliver sharp, contrasty nice images, and at $2700 including a very solid tripod and mount, a good deal less expensive than an apochromatic refractor of comparable aperture.

Celestron also offer a small go-to Maksutov, the NexStar 4, similar in specification to the ETX 105 but notably cheaper at around $500. Unlike the ETX telescopes, the NexStar 4 operates reasonably well without a tripod or mount, resting on a tabletop. It has rubberized feet to provide the necessary stability, but a tripod does make life a lot more pleasant by raising the telescope to a comfortable observing height. A lightweight "NexStar 4/5" mount and tripod ($170) is available and designed to be operate in both alt-azimuthal and equatorial mode. With the tripod in equatorial mode and the telescope tilted on the supplied wedge, the NexStar 4 is particularly well suited for astrophotography. Playing on this strength, Celestron have equipped the NexStar 4 with a built-in flip mirror that allows the user to attach both a camera and an eyepiece to the telescope at the same time. You can then toggle between looking through the eyepiece for finding the target and taking the photograph with the camera. Like all the NexStar telescopes, the NexStar 4 comes with a simple red-spot reflex finder. Views through the NexStar 4 are good, allowing for the limited aperture, though

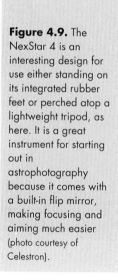

Figure 4.9. The NexStar 4 is an interesting design for use either standing on its integrated rubber feet or perched atop a lightweight tripod, as here. It is a great instrument for starting out in astrophotography because it comes with a built-in flip mirror, making focusing and aiming much easier (photo courtesy of Celestron).

perhaps not quite as good as those delivered by the ETX 105. In addition, the star diagonal and eyepiece that come with the ETX are distinctly better than those with that come with the NexStar. On the other hand, the NexStar 4 is a good deal less expensive, and couple that with the fact it works without a tripod and the NexStar 4 has a lot to recommend itself to beginners wanting a small and portable go-to telescope.

An SCT instead of a Maksutov, the NexStar 5i ($900 + $150 for the go-to handset) uses one of Celestron's most highly respected optical tubes. Over the years, their 125-mm (5-inch) SCT tube has been included in a variety of designs, and has consistently delivered good quality images. Compared with the ETX 125, the NexStar 5i performs very similarly, but the focal length is shorter and so for any given eyepiece will deliver a lower magnification and wider field. One thing to remember about SCT optical tubes is that they do require careful collimation to perform well, whereas Maksutov designs (like the ETX 125) do not. Of the two designs, Maksutovs generally get the nod for planetary and lunar observing thanks to their near-refractor performance, whereas the wider field of the SCT offers the best all-round, solar system and deep sky observing. The NexStar 5i is certainly a very handsome telescope, with a cast aluminum tube mounted by a single-arm (but sturdy) onto a sturdy base with rubberized feet similar to those seen on the NexStar 4. Like the NexStar 4, this telescope will work resting on a table or some other firm surface, but the NexStar 4/5 mount and tripod ($170) is a good investment nonetheless. Unlike the other go-to telescopes mentioned in this chapter so far, the go-to part of the package (the Autostar handset) isn't included as standard equipment but as a $149 upgrade. Out of the box, the NexStar 5i, and its larger brother, the NexStar 8i, use a non-go-to handset with traditional push button controls for slewing and so on.

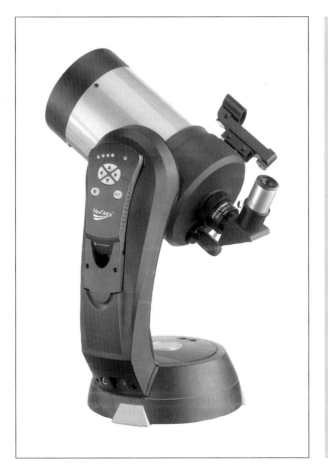

Figure 4.10. For its aperture size, the NexStar 5i is relatively expensive, but you do get one of the best mass-produced SCT optical tubes around in a nicely designed package (photo courtesy of Celestron).

Table 4.4. Summary of Maksutov and smaller SCT features

Pros:	Generally, the most compact instruments, and stable on even relatively modest tripods. Small size makes these telescopes much more portable than Newtonians of similar aperture, while offering more aperture than a refractor of similar cost. No collimation required for Maksutovs, and compared with Newtonians, collimating an SCT is easy. Images without false colour.
Cons:	Of all the common designs, SCTs deliver the least contrasty images, the detail is still there, but colours tend to look a bit more muted. Closed design traps heat inside the optical tube, and these telescopes can take an hour or more to cool down. Long focal lengths (speeds of f/10 or more) not suited to wide-field observing. Aperture too small for effective use of light pollution filters, particularly narrow-band ones.
Ideal for:	Most observers wanting a compact, versatile telescope will find these telescopes a good compromise between size, cost and optical quality. Suburban astronomers limited to solar system and lunar observing will appreciate the excellent resolution and contrast of Maksutovs in particular.
Not ideal for:	Budget-conscious observers will find Newtonians offering a better price-to-performance ratio, though with diminished portability and convenience. Astronomers most interested in lunar and planetary observing may find the views through SCTs lacking in contrast compared with refractors and long focal length Newtonians.

200-mm (8-inch) and Larger SCTs

The 200-mm (8-inch) SCT has become the *de facto* standard telescope for amateurs because of its excellent balance between cost, performance, portability and
size. It has a large enough aperture to perform well on deep sky and solar system
targets, and on nights of average seeing will allow about as much resolution as
you can hope to attain as far things like lunar detail goes. Refractors of comparable aperture are frighteningly expensive and demand a permanent installation;
and although a Newtonian is much cheaper than a similar sized SCT, they are big,
unwieldy instruments, particularly when equatorially mounted. Many amateur
astronomy books and magazines assume an aperture of 200 mm or more when
describing the appearance of deep sky objects, and so a 200-mm SCT is a versatile
instrument for the hobbyist without a specific interest but just wanting to look at
a little bit of everything. The SCT design enjoys many different upgrades and
accessories too. Focal length reducers (or reducer–correctors) are particularly
popular, converting an f/10 telescope into an f/6.3 one, and most SCT tubes come
with special holes for clipping on cameras, counterweights, even small telescopes.
If the apochromatic refractor represents a Zen level of purity in astronomy, then
the SCT is the telescope of choice for the techno-junkie! There are a few shortcomings though. Large SCTs take a long time to reach thermal equilibrium (as
much as two hours) and the view through the eyepiece is not quite as sharp as
with a refractor or long focal length Newtonian. Mirror shift when focusing is

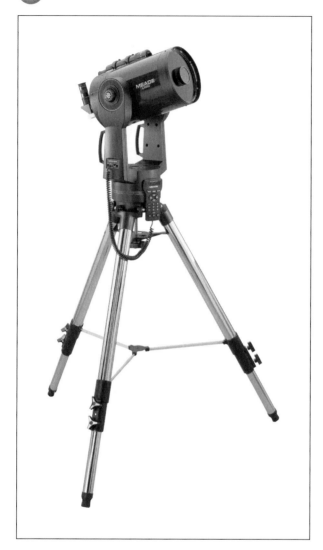

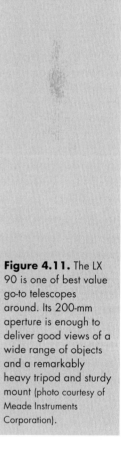

Figure 4.11. The LX 90 is one of best value go-to telescopes around. Its 200-mm aperture is enough to deliver good views of a wide range of objects and a remarkably heavy tripod and sturdy mount (photo courtesy of Meade Instruments Corporation).

very annoying, and at high powers can make focusing very difficult if the target keeps bouncing out of view.

These problems are minor ones though, and both Meade and Celestron offer wide ranges of go-to equipped SCTs with 200 mm of aperture or more. Meade's line splits into three: the budget-priced, equatorially mounted LXD 55; a mid-priced fork-mounted model, the LX 90; and a series of fork and pier mounted models aimed at advanced observers, particularly astrophotographers, the LX 200 series. The LXD 55 SCT ($1400) complements the LXD 55 Schmidt–Newtonians discussed earlier, offering astronomers on a budget something of the utility of the higher end instruments, but like those telescopes, it suffers from having a light tripod and weak mount. The LX 90 is a very different beast altogether, coming on the same solid tripod as the LX 200 telescopes, but with the optical tube held by a smaller, but still cast metal, fork. Though a bit more expensive ($1700) than the LXD 55, this is one of those instances where paying the extra makes a real difference. For amateurs interested in both observing and basic astrophotography (for example with a webcam), the LX 90 has proved to be a versatile and popular choice. Its only real shortcomings are that the fork arms are too short to allow a mass of photography equipment to fit between them and the mount, and the lack of periodic error correction to smooth out irregularities in the motors. Both these flaws are absent from the LX 200 SCTs, which look like beefier versions of the LX 90. The much larger forks allow the back end of the telescope to hold all sorts of astrophotography gear and still leave enough clearance when the telescope points upwards. The LX 200 telescopes also come with a host of other small but useful features such as electric focusers and additional power ports for accessories and. They also possess a global positioning satellite (or GPS) receiver for speeding up the alignment process. GPS isn't an essential feature (it doesn't offer anything you

Figure 4.12. The LX 200 telescopes make excellent instrument as for visual observing, but their real forte is as a rig for long exposure astrophotography. Heavy-duty mounts and strong fork arms allow these telescopes to bear all manner of equipment without complaint (photo courtesy of Meade Instruments Corporation).

can't do quickly yourself) but many people have found it to be very convenient, easy to use and reliable. LX 200 telescopes run from $2000 for the 200-mm tripod-mounted model through to $16000 for the 400-mm pier-mounted one.

Celestron offer the NexStar 8i ($1200) at as their entry-level 200-mm SCT. Essentially a scaled up version of the NexStar 5i mentioned earlier, one of the most notable differences between it and the Meade range is the use of a carbon fibre optical tube instead of a metal one. This speeds up cooling somewhat, and reduces the weight of the instrument making it much easier to carry about. Like the NexStar 5i, the NexStar 8i has a single arm supporting the optical tube rather than a two-armed fork like most other SCT designs; generally this works well for visual observing but can be shaky at high powers and especially when used for astrophotography. Like the NexStar 5i, the go-to handset is an optional extra, though a lightweight tubular steel tripod is included. Even after adding the handset, at $149, the NexStar 8i is quite a bit less expensive than the Meade LX 90.

Figure 4.13. The NexStar 8 GPS is easily the sleekest and most high-tech looking of the larger SCTs. A key innovation is a carbon fiber tube that reduces weight and allows the telescope to cool down more quickly than traditional optical tubes (photo courtesy of Celestron).

So how do you choose between the two? The NexStar 8i is lighter and easier to set up and transport, but the single-arm design and lightweight tripod put it at a disadvantage as far as stability is concerned. In contrast, the two-armed cast aluminum fork and much heavier tripod make the LX 90 about as close to rock solid as you can expect from a mid-priced telescope. A step up in price is the NexStar 8 GPS ($2000) which does have two-armed fork and a sturdier tripod, as well as coming with the go-to handset as part of the package. The NexStar 8 is above the Meade LX 90 in specification, having permanent periodic error correction for better long exposure astrophotography, and GPS positioning to make the

Figure 4.14.
Celestron produces a number of Advanced Series go-to SCTs on equatorial mounts, such as the C9¼S-GT (photo courtesy of Celestron).

Table 4.5. Summary of large SCT features

Pros:	Excellent compromise between cost, portability, convenience and aperture. Minimal maintenance required. There are many suitable accessories for this design, including reducer/corrector lenses, light pollution filters that fit on the telescope rather than the eyepiece, and so on.
Cons:	Images noticeably less colorful or sharp as those from the other common designs, and optical tubes take a long while to cool down. At the cheaper end of the market, the tripods, mounts and other accessories are often unimpressive. Narrow field of view of the f/10 models not ideal for wide-field observing, though reducer/corrector lens can help somewhat.
Ideal for:	Any observer wanting a versatile instrument capable of delivering satisfying views of most anything. Astrophotographers.
Not ideal for:	Observers after an ultra-portable scope or a telescope offering the very sharpest, most contrasty images will be better served by a refractor of some type. Budget-conscious amateurs will get better value from a Newtonian.

alignment procedure easier. Similarly appointed NexStar telescopes of 235-mm (9.25-inch) and 280-mm (11-inch) aperture sizes are also available, priced at $2700 and $3000 respectively. Celestron's Advanced Series of equatorially mounted telescopes feature two SCT designs, in each case on CG-5 mounts and tubular steel tripods and employing the NexStar go-to system. These are roughly equivalent to the Meade LXD 55 SCT, although the mounts are substantially sturdier. The 200-mm C8S-GT ($1300) is perhaps the better of the two, the 235-mm C9$1/_4$S-GT ($1700) being a bit too big and heavy to really work well with the CG-5 mount. Celestron also produce a CGE Series of high-end equatorially mounted go-to telescopes from 200 mm to 350 mm (8 to 14 inches) priced from $3500 through to around $6000. These are primarily suited to advanced amateurs, particularly astrophotographers, who prefer equatorial mounts to the fork mounts used by the Meade SCTs.

Bespoke Go-To Telescopes

Although the selection of go-to telescopes offered by Celestron and Meade is broad, the SCT design dominates, particularly among the intermediate and high-end models. This is fine for observers with general interests, but specialists often favor particular designs, such as apochromatic refractors or classical Cassegrains that have advantages when used for certain tasks. There are three ways of bringing go-to functionality to the optical tube of your choice: sitting it piggyback on a go-to SCT; grafting it onto an all-in-on mount with the original optical tube removed; or turning a normal equatorial mount into a go-to one. The following section describes these methods, each having its plusses and minuses, and Internet links to the manufacturers discussed are in Appendix 1.

Go-To, Piggyback Style

This is the simplest and lowest cost way to add go-to functionality to a regular telescope, but it is the most limited as well. Celestron, Losmandy, Meade and ScopeStuff, among others, market a range of attachment plates and counterweights that fit onto the 200-mm and larger go-to SCT telescopes from Celestron and Meade. These allow the user to take a traditional non-go-to telescope such as small refractor and "piggyback" it onto the SCT rather like a super-sized finderscope. If the smaller telescope is a wide-field instrument, then having the chance to get two different views of the same deep sky object can be enchanting; the Milky Way clusters in constellations like Auriga and Sagittarius, for example, would be very rewarding studied in this way. Alternatively, a piggybacked apochromatic refractor could take advantage of the SCT's go-to ability and computerized tracking to give the observer great views of the planets, while the SCT itself would handle deep sky duty where aperture is the most important thing. If you do long-exposure astrophotography, a piggybacked telescope would be a useful guidescope, too.

All piggyback kits work in the same way: a rail or plate attaches to SCT optical tube using small screws, and the small telescope locks onto this. You don't need to drill anything, all modern SCT tubes come with special holes of attaching things built in; if you haven't tampered with your SCT yet, then the holes will be occupied by small screws that wind out using an Allen key. Onto this rail go a pair of cradle rings, and these need to be wider than the telescope that they are going to hold. Normally these rings hinge open so that the telescope lifts in an out

Figure 4.15. Notably missing from the Meade and Celestron range of go-to telescopes are top-quality apochromatic refractors such as this TV 76. Since these telescopes are commonly bought to supplement a larger "light bucket" SCT, several manufacturers produce kits to attach them onto a go-to SCT effectively producing two computerized telescopes for the price of one (photo courtesy of Tele Vue Optics, Suffern, NJ).

Table 4.6. Piggyback go-to telescope features

Pros:	The required parts are inexpensive and easy to assemble.
Cons:	Only an option for mounting small telescopes, primarily 80 mm and smaller aperture refractors, onto large 200 mm and bigger SCTs.
Ideal for:	Amateur astronomers with a computerized SCT and a quality wide-field refractor wanting to get the different views these telescopes offer at the same time.
Not ideal for:	Any observer that doesn't have the combination required or wanting such a set-up.

easily. If your SCT is equatorially mounted, you may be able to get the whole thing to balance using the counterweight, but alt-azimuthal SCTs (such as the LX 200 and NexStar 8) don't have counterweights, and the additional weight of the piggybacked telescope will make it impossible to balance the telescope properly. Tightening the clutched on the fork arms is not an option; too much strain here can damage the motors and will certainly mess up go-to accuracy. Instead, you will need to add a telescope balancing kit, essentially a rail with sliding counter-weights, underneath the telescope, again using pre-existing holes in the optical tube. Once you've done this, you can move the weights backwards and forwards until the telescope balances correctly. You may even need to add extra weights, sold separately, for very heavy loads. You'll need to align the piggybacked tele-scope with the main telescope in the same way as a finderscope, using the three screws on each of the cradle rings. With that done, both the main and the piggy-backed telescope will work together beautifully; and for a modest outlay – about $160 for the rings, rails and counterweights – you'll have a second go-to telescope!

Hacking an "Off the Shelf" Go-To Telescope

If you have a refractor or small compound telescope but no large SCT to mount it onto, one possible option is to recycle a cheap go-to telescope, replacing the optical tube that came with it with one of yours. Since even a "cheap" go-to tele-scope costs a fair bit of money, this option only makes sense if your existing tele-scope is of sufficient quality to justify what is essentially buying another telescope, throwing away the optical tube it came with, and voiding the warranty to boot! Of course, you might be able to sell the unwanted optical tube to another amateur, perhaps for use as guidescope, or else you can buy the go-to telescope second-hand and save some money that way. One of the most popular hacks is to replace the Maksutov tube in the ETX 90 with the Tele Vue Ranger wide-field refractor. These two telescopes have very different strengths, the Maksutov for high-power views of the planets, and the refractor for wide field views of open star clusters, so swapping the tubes one for the other creates a very different observing tool. This approach is rather trial-and-error as well, as not all combina-tions of optical tube and go-to telescopes work, and some skill and ingenuity will be required to make the modifications.

If you'd sooner by something in kit form or ready made, then look to the German astronomical equipment manufacturer, Baader Planetarium. They produce conversions of go-to mounts taken from the Celestron 60, 80 and 114 GT go-to telescopes. Included in the package are a dovetail clamp and a dovetail bar. The dovetail bar slides in and out of the clamp, and holds the replacement optical tube. Machined be compatible with the Vixen refractors and their clones, such as those from Orion and Celestron, the dovetail bar slides into the space between the jaws of the dovetail clamp making it very easy to fix the telescopes into place. For designs lacking this standard Vixen fitting, the Baader dovetail plate also accommodates cradle rings similar to those used for piggybacking telescopes onto go-to SCTs. The modified Celestron go-to mount will support anything up to about 4.5 kg (10 lbs) successfully. Being small and portable, these alt-azimuthal mounts work especially well with compact instruments such as the smaller Maksutovs and short focus refractors. As such, they can provide a useful platform for owners of small, deluxe instruments like the Questar 90-mm (3.5-inch) Maksutov and Takahashi refractors, or spotting scopes such as those from Leica and Zeiss not normally used as astronomical instruments but of sufficient optical quality to warrant the expensive of a go-to mount. Ready-made go-to mounts of this type cost $560; otherwise, the kit for hacking a user-supplied Celestron go-to telescope is $70.

Go-To Mounts and Upgrades for Traditional Mounts

Piggybacking a small refractor onto a large SCT or replacing the optical tube on an existing go-to telescope are viable enough ways to custom-build a go-to telescope, but they are both limited to being upgrades for small optical tubes. For observers wanting to upgrade larger telescopes, particularly designs that don't come pre-packaged as go-to telescopes from either Meade or Celestron, the best solution is to use a go-to equipped mount. Several companies produce such mounts, in the case of Vixen, for example, to complement their own line of traditional (non-go-to) telescopes. One prime advantage a go-to mount has over an all-in-one go-to telescope is that it is a flexible system: you can change the optical tube as the situation demands, perhaps using a refractor for looking at double

Table 4.7. Rebuilt go-to telescope features

Pros:	Easy to put together, and even easier if bought pre-assembled. Effective way to computerize a small telescope.
Cons:	Expensive and limited to only relatively small telescopes.
Ideal for:	Observers with top-quality small refractors or Maksutovs that justify the expense of breaking down a perfectly serviceable commercially produced go-to telescope for the required parts.
Not ideal for:	Observers on a budget or with a telescope design too large for the small mountings for which this method works.

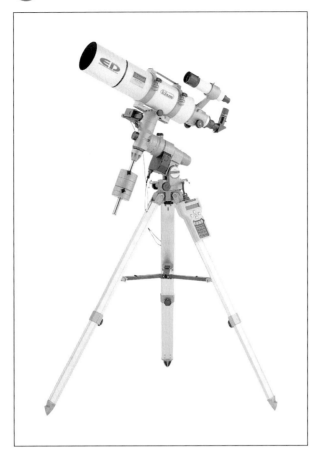

Figure 4.16.
Computerized mounts, such as this SkySensor unit mated to a premium refractor, are never cheap, but they are by far the most versatile of all the go-to solutions available to hobbyists. The Sky Sensor system includes the handset, motors and cables required to convert a traditional equatorial mount to full go-to capability (photo courtesy of Vixen North America).

stars on one night, and a fast Newtonian for deep sky work the next. These are also systems to grow with; as your skill and interest develops, you can sell your original optical tube and upgrade to a better one. In operation, these mounts work in much the same way as the Autostar and NexStar systems, but without the liabilities of the lightweight aluminum mounts supplied with many of those systems. There is really only one drawback to a go-to mount, and that is that these mounts tend to be relatively expensive, having been designed more for performance and durability than price: many of these mounts cost more than a complete go-to telescope from Celestron or Meade. Smoother, more precise drives, metal rather than plastic components, and machined steel legs instead of extruded aluminum ones all add to the price but improve the stability, tracking and accuracy of the mount.

Vixen produce two go-to systems, an upgrade called SkySensor 2000 for their traditional mounts, and an all-in-on computerized mount known as Sphinx, onto which can be mounted any compatible optical tube. The SkySensor package ($1200) includes the handset and all the cables and motors required (which replace any existing motors). It fits into place quite easily, and although designed

Figure 4.17. Vixen's Sphinx go-to system combines the flexibility of a go-to upgrade with the all-in-one convenience of a typical go-to telescope. All the encoders and cables are tucked away inside the mount and tripod, and the handset is particularly appealing, featuring a graphical planetarium program (photo courtesy of Vixen North America).

for Vixen mounts, the SkySensor is remarkably adaptable, and some hobbyists have been able to use it with the Chinese and Taiwanese mounts that are clones of the Vixen ones, for example many of the mounts that come with Celestron and Orion telescopes. The Sphinx system (from $1800) is a more sophisticated system with all the gears and motors hidden inside the mount. Key features include better tracking when compared to an upgraded Vixen Great Polaris mount and lower energy consumption so that batteries last longer, but the most obvious difference is in the use of a totally different handset. In fact, the Sphinx handset, known as Star Book, is quite unlike any other computer handset, featuring what amounts to a built-in planetarium program, and makes a nice stand-alone sky atlas in its own right. When used with small optical tubes, like the Tele Vue Pronto, the Sphinx mount works fine with an easily to carry tabletop tripod ($115), but otherwise an aluminum tripod ($340) rated for optical tubes up to 9.9 kg (22 lbs) should be used for maximum stability.

Made primarily from machined steel rather than plastic and aluminum, the Losmandy mounts and tripods are a step up from Vixen's in terms of price and quality and are most popular with astrophotographers and advanced hobbyists demanding smooth and steady, tracking for long exposures and high-magnification observing. Like the SkySensor 2000, the Gemini go-to system ($1600) is an upgrade for their traditional mounts, the GM-8, G-11, HGM200, CI-700 and G9. One of the neat features of the Gemini system is that it is compatible with a Windows CE compatible Pocket PC running *TheSky*. A palmtop computer like this is no bigger than the handset of a typical go-to system and gives the Gemini system a graphical interface similar to that offered by a laptop computer but in a much smaller package. Long famous for delivering some of the very best apochromatic refractors around, Astro-Physics also produce top-quality go-to mounts. The GTO series runs from the 400 GTO, which can carry up to 8.2 kg

Table 4.8. Go-to mount features

Pros:	Versatile, work with practically any telescope, and the optical tube can be removed and replaced with another as necessary. Generally, the mountings are of the equatorial design and much sturdier than the lightweight aluminum mountings that come with many entry and mid-level telescopes, and therefore ideal for astrophotography. Build quality excellent.
Cons:	Relatively expensive, and in the equatorial format often heavy, cumbersome and complicated.
Ideal for:	Advanced amateurs with a number of optical tubes used for different targets and on different occasions. Sturdy equatorial mounts are particularly useful for astrophotographers.
Not ideal for:	Observers wanting a portable, low-cost, all-in-one package.

(18 lbs), through to the 1200 GTO rated for 63.6 kg (140 lb) loads. None are cheap – the 400 GTO sells for $3500 and the 1200 GTO for $7500 – but like the Losmandy mounts these are about as good as they get, and widely respected by advanced amateur astronomers. Another player at the top end of the market is Takahashi with their go-to versions of their popular Temma series of equatorial mounts. Unlike the other mounts listed here, the Temma lacks its own hand controller, and needs a compatible PC running *TheSky* software (note though that the required software won't work with either the Mac or Windows CE versions of *TheSky*). The Temma go-to mounts range from $3700 for the 7 kg (15.4 lbs) capable EM-10 through to the $15000 EM-500 mount that can carry over 40 kg (89 lbs).

Figure 4.18. The Sky Tour upgrade doesn't turn a telescope on a Tele Vue mount into a go-to telescope, but it does make it easier for the user to manually point the telescope at the desired target (photo courtesy of Tele Vue Optics, Suffern, NJ).

Digital Setting Circles and Go-To Dobsonians

Digital setting circles, or DSCs, such as Meade's Magellan system, essentially give the user and easier way to read the setting circles and combine this with a database of deep sky and solar system objects. In contrast to go-to systems, DSCs don't move the telescope, the user must still do that, instead they display the coordinates very accurately in a liquid crystal display, making it simpler to ascertain how much and in which direction the telescope needs to be moved. Once centered, the motors track the object normally, if the telescope has such motors. Compared with the analogue setting circles (i.e., those engraved onto the mount itself), which many people found completely baffling, DSCs were a significant step forward in terms of clarity and ease of use (not least of all because the DSC screen displays the exact RA and declination values in illuminated, easy to read numbers). DSCs can be used in the other direction too: manually find and point the telescope at something, and the DSC will at least tell you the right ascension and declination, and if there is an object in its catalogs, what that object is.

Frankly, DSCs never really caught on. Early examples were expensive, fiddly to install and calibrate, and many users found them more trouble than they were worth. By the time the technology did become reliable and inexpensive, fully automated go-to telescopes were available, and these made much more sense for the average amateur. A few companies still produce them though, primarily for telescopes on mounts that don't lend themselves to go-to computerization, for example Tele Vue produce the Sky Tour DSC system for use on their portable alt-azimuthally mounted refractors. Similarly, giant Dobsonian telescopes are by far the very best instruments for tracking down deep sky objects, and DSCs provide a convenient way to track down faint and obscure targets. Among the manufacturers of DSCs for Dobsonian telescopes are Jim's Mobile, Lumicon, Orion and Sky Commander. There are even kits available that turn the DSCs into a go-to system, by adding on the necessary motors, such as that from Tech2000. The StarMaster

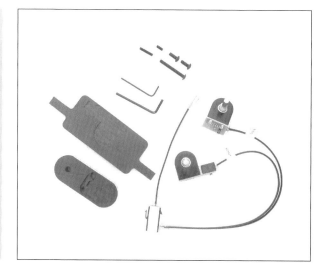

Figure 4.19.
Upgrading a traditional mount to either digital setting circle mode, as with this Sky Tour kit, or to full go-to capability, requires adding encoders and wires and, in the case of go-to upgrades, motors as well (photo courtesy of Tele Vue Optics, Suffern, NJ).

Table 4.9. Digital setting circle features

Pros:	Versatile and can be used with many different telescopes, including alt-azimuthally mounted telescopes such as Dobsonians. Some DSCs are reasonably inexpensive.
Cons:	Don't offer go-to and require the user manoeuvre the telescope into position (though go-to may be available as an upgrade). Can be expensive and fiddly to use.
Ideal for:	DSCs are ideal for observers wanting to apply computerized telescope functionality to a "light bucket" Dobsonian or other manually operated telescope.
Not ideal for:	Observers wanting an all-in-one package.

open-truss Dobsonians are available as complete go-to systems, though these are deluxe optics featuring some of the best optics around: go-to is a $2200 upgrade to instruments that already cost several thousand dollars.

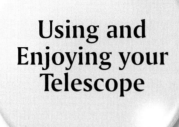

CHAPTER FIVE

Using and Enjoying your Telescope

Having looked at the various kinds of go-to telescopes, the next thing to do is find out how to use it best. There are lots of books with lists of things to point a telescope at and discussions on the accessories you can buy to customize it to best fit your needs; rather than go over all that, this chapter is specifically about ways to improve a go-to telescope. Some of the materials covered in this chapter apply equally well to non-go-to telescopes, for example, methods for improving light-weight tripods and selecting the right sorts of eyepieces, so even if you don't have a go-to telescope this chapter should still be of value. But a lot of what is here is directed at users of the Autostar and NexStar telescopes from Meade and Celestron, or any of the various go-to mounts produced by other companies.

Making the Most of Lightweight Tripods and Mounts

Many of the telescopes described here come with tripods and mounts, usually made from extruded aluminum with plastic fittings. Although durable and strong, aluminum suffers in comparison to wood or steel from being rather light, and the aluminum legs in particular transmit vibrations from the ground splen-didly well up to the telescope. The result is that while the telescope may seem sturdy enough, when you are looking at objects under high magnification, the views will seem to jump about. This is not only annoying but makes things like focusing and taking photographs incredibly difficult. Of course, the best thing you can do is to make sure you get a telescope on a heavy mount and sturdy tripod; but is there anyway to redeem a go-to package that comes with light-weight tripod and mount? Up to a point, the answer is yes.

Aluminum is very strong, for its weight stronger than most kinds of steel, and in fact, the typical aluminum tripod will support much more weight than that of the telescope. The problem with aluminum isn't its lack of mechanical strength in compression; it has plenty of that. No, the problem with aluminum tripods is that they are much lighter than the loads they carry: the telescope, its counterweights and perhaps accessories like jumbo eyepieces and cameras. Consequently, the entire structure becomes top heavy, and this has the effect of magnifying vibrations: in other words, it wobbles. To damp down the vibrations you need to bring the center of gravity down by making the lower part of the structure heavier. One way to do this is to fill the legs with clean sand. Silica sand, also known as silver sand, is ideal as it tends not to absorb moisture in the same way as beach sand, and is chemically inert. Garden centers and horticultural supply stores sell silica sand for use with houseplants, particularly cacti. Usually each leg of the tripod is in two sections, a narrow bore lower section, and a wide upper section into which the lower section can slide. Carefully remove the lower sections. These are the bits that can be filled with sand; put the rest of the tripod aside for the time being taking care not to lose any screws and bolts that had to be removed to get this far. Depending on the exact construction of your tripod there will be plastic or rubber feet at the base of these lower sections and plugs at the top. Remove the plugs but leave the feet in place. You should now be able to see into the hollow space inside the leg, and by looking to see where light gets in, you can establish easily where sand would leak out. Use non-toxic silicone sealant (from your local tropical fish retailer) to glue up any holes around screws and bolts along the legs, and the gaps where the rubber or plastic feet join onto the aluminum. Silicone sealant is easy to scrape or pick off if you need to, but is otherwise strong, non-toxic once dried, and waterproof. Silicone sealant is a skin irritant though while wet, and care taken to avoid prolonged contact and to follow any handling guidelines listed by the manufacturer. The acidic vapours it gives off while drying smell nasty too, so do this job outdoors! The sealant sets in about an hour, and fully cured about a day later. Once the aluminum legs are sealed, pour sand in carefully and then put the various plugs and screws back into place. Seal these with sealant if you want. Slide the sand-filled lower sections back into the upper sections, and re-assemble

Figure 5.1. Various companies produce vibration dampeners that try to mitigate the tendency of lightweight tripods to shake when touched (photo courtesy of Celestron).

the tripod. Once the tripod and mount fitted back together, you will notice that the entire structure is much heavier than before and a good deal sturdier.

If it still isn't stable, you can add more weight. Try suspending a bucket of sand from the tripod. It needs to be hanging low, but mustn't touch the ground. Ice cream cartons are an ideal size for this. The aim is to get the center of gravity of the whole assembly as low down as possible; the lower it is, the more it will resist any tendency for the optical tube to wobble. A final boost to the usefulness of the tripod comes from using shock absorbers under the feet of the tripod. A number of astronomy companies make their own versions of these, including Meade and Celestron, costing around $50 for a set of three, but many amateurs have pressed into service all manner of alternatives. The aim is to make something that will absorb vibrations in the tripod between the feet and the hard surface the tripod is set up on. Anything made of dense fabric or rubber works well for this, such as bits of carpet underlay, bathtub plugs and even small pieces of old tyres.

Go-To Alt-Azimuth Versus Equatorial Mounts: Which Is Best for You?

Telescope mountings come in two main designs, the *alt-azimuth and* the *equatorial.* Alt-azimuth mounts, such as the fork mounts such as those employed by the NexStar and LX series telescopes, have two axes of movement at ninety degrees to one another, side to side (azimuth) and up and down (altitude). Although computers can control an alt-azimuthal mount sufficiently well to allow it to track objects, they cannot prevent something called *field rotation*. This comes about because of the fact that astronomical objects are not travelling across the sky in a straight line, but going around the celestial poles in great circles. (Of course, it is the Earth and everything on it that is rotating, but we'll ignore this for now.) A computerized alt-azimuth mount follows an object as it rises in the east, across the sky, and then sets in the west. If you look down telescope, though the object will stay in the field of view, it will start off seeming to "lean" in one direction as it rises, straighten up as it reaches its highest point in the sky, and then tips over and leans in the other directions as it starts to set. This doesn't matter for short-exposure photography (such as when using a webcam or digital camera for simple snaps of the Moon and planets). However, it does make a difference if the exposure lasts for more than a few minutes, when things start to get blurry. To do long-exposure astrophotography you need to be able to compensate for this apparent rotation of the field of view. A *field de-rotator* will turn the camera (or CCD) in the opposite direction to the rotation of the field of view, and so cancel out the rotation of the field over time. In practice, these devices work adequately well but many astrophotographers find it more effective to convert an alt-azimuthal mount into an equatorial one by adding an *equatorial wedge* to the tripod. This only works with compact instruments like SCTs and Maksutovs that

Figure 5.2. A good equatorial mount is probably the best all-round choice for the amateur astronomer; the only shortcoming of this design is its size and weight (photo courtesy of Celestron).

are small and light enough not to cause the now-tilted mount to become unstable. For other designs, a proper equatorial mount is essential, such as those employed by Celestron's Advanced Series.

Like alt-azimuthal mounts, equatorial mounts have two axes of movement, again at ninety degrees to each other, but instead of being horizontally and vertically aligned the two axes are at an angle to the Earth's surface. This angle depends on the latitude of the observer. The advantage of this is that if the right ascension axis is pointing at the northern or southern celestial poles, depending on the hemisphere of the observer, the a slow but constant rotation of the telescope around the right ascension axis will allow the telescope to follow a star or planet as it moves across the sky. Equatorially mounted telescopes automatically compensate for field rotation because instead of following the object in altitude and azimuth, they are actually rotating under the celestial pole around the right ascension axis, with the result that the telescope is rotating at the same rate as the object it is pointing at. The upshot of this is whatever the telescope or camera is pointing at holds its position in the field of view. There are still other things to worry about, such as periodic error correction and precise polar alignment, and these fall outside the scope of this book. However, if you plan to do astrophotography seriously, then it is just as well to appreciate the limitations of the alt-azimuthal mount before buying one.

Aligning a Go-To Telescope

Having bought and unpacked your telescope, you now need to wait for inevitable post-astronomical purchase rainstorms to pass. These usually last about a week and will let up just in time for Full Moon when you won't be able to see much of anything other than the Earth's natural satellite. While you're waiting for the clear, dark skies to return, it's time to read the manual that came with your telescope. One of the topics it will concentrate on is alignment. Getting this right is essential to getting the best performance from a go-to telescope. You're manual will tell you the specifics for the system you have, but most work in the same basic way, by triangulating the position of the telescope from measuring the angles between three points, normally due north (or south) plus two bright stars. Since the stars for all practical purposes have fixed positions in the sky, the angular distance between any two stars is a constant; but the angle between these stars and due north (or south) will of course vary, but in a predictable way, which is why your telescope will require the date and time to work out these angles.

As the exact method for setting the telescope pointing northwards (or southwards) and then slewing to the two bright stars will vary with whether you use Autostar, NexStar or one of the go-to mounts, there isn't much point citing them all here. What is worth stating are some useful tips for improving the alignment process, and if you bear them in mind as you go along, you will minimize the likelihood of poor go-to performance and inaccurate tracking. These considerations are particularly important to astrophotographers, but even beginners will find life much simpler if the telescope manages to center targets in the field of view reliably every time. By the way, GPS simplifies some of this process, by giving the computer accurate time, date and location data; but centring the field of view on the two bright stars is still required. DSC systems are set up in the same sort of way as a non-GPS go-to system so much of what follows here goes for them as well, except of course you manually push the telescope or work the motors to find the alignment stars rather than wait for the computer to automatically slew to them.

Use the Most Accurate Time You Can

This makes a huge difference because the telescope achieves pretty much everything it does in a time-dependant way, and so any errors in the time that it is given to start with will wind up reducing the pointing accuracy profoundly. Many astronomers like to use electric clocks that use military time signals, while others will take their time from an Internet time server; both these methods ensure that you begin with the most accurate time possible reducing that particular source of

Figure 5.3. A go-to telescope will perform properly if the mount is aligned properly and computer handset, such as this Autostar unit, is given accurate information to work with (photo courtesy of Meade Instruments Corporation).

error. If you take a laptop into the field with you, then synchronize its clock to a network time server on the Internet, and you'll have a handy and very accurate clock at your disposal. GPS-equipped telescopes should be able do this step for you.

Figure 5.4. Many telescopes are able to obtain accurate time and position data from the global positioning satellite system, either using built-in receivers or via upgrades, such as the GPS Accessory CN16 for the NexStar 5i and NexStar 8i telescopes (photo courtesy of Celestron).

Use Accurate Longitude and Latitude

As with time, the go-to computer uses this information to work out the locations of the bright stars it will slew to during the alignment process. Since 90° minus your latitude is equal to the altitude of the northern celestial pole (in degrees) above the northern horizon, or the southern celestial pole above the southern horizon. Where do you find your longitude and latitude? A good gazetteer should provide this information (check your public library if you don't have one); there's one in the *Encyclopaedia Britannica*, for example. Airports and harbours also use this information, and if there is one nearby, you could try calling them to find out. Handheld GPS receivers are popular with hikers and other country sport enthusiasts, and these will give you very accurate readings of longitude and latitude; and as with date and time, GPS-equipped telescopes are able to get this bit of information by themselves. You'll need degrees and minutes for both, and if you get your longitude and latitude in degrees, minutes and seconds, you'll need to round it up or down, so don't forget that less than thirty seconds goes *down* to the nearest minute, and thirty seconds or more *upwards*. Moving your telescope around your back garden won't demand an update to the longitude and latitude, but if you drive out to a star party, you may well need to. A degree latitude corresponds to about

111 km (69 miles) wherever you are, but the distance between lines of longitude changes, starting at 111 km apart at the equator converging to zero at the poles.

Level the Telescope as Best as Possible

The go-to procedure usually begins with the optical tube in a horizontal position and pointing due north (or south, depending on where you are). Remember, this isn't magnetic north (or south), so a compass isn't much good for this; instead rely on identifying the relevant pole star for your hemisphere. Although in theory the computer should be able to compensate for the tube being not quite horizontal at the beginning, which could easily be the case if the tripod is set up on uneven ground, many users have found the go-to accuracy to be better if this source of error is minimized to begin with. The best way to do this is to ensure the telescope is horizontal by using a spirit level of some sort. Particularly nice are the eyepiece-sized ones produced by Broadhurst, Clarkson and Fuller for about $35. These slip into the eyepiece holder or star diagonal and indicate when the tube is level; they also have a magnetic compass built into them too, which as mentioned before isn't much use for aligning the telescope. Alternatively, a regular woodworker's spirit level can be used to check a tripod is level before setting the telescope onto it, and then once the telescope is on the tripod the spirit level can be rested on the optical tube to double check it is horizontal. Of course, this won't work if the optical tube tapers, but luckily, most reflectors and catadioptrics don't, but this may be an issue with some refractors. Once again, this is a bit of the alignment procedure that GPS systems simplify, being able to level themselves automatically.

Use High-Magnification or Reticule Eyepieces During Alignment

Many go-to telescopes come with relatively low-magnification eyepieces that are fine for observing with but not the best things to use when aligning. Alignment depends on the user positioning the "alignment stars" dead center, and the larger the field of view the less accurate this estimate is likely to be, and the greater the resulting errors in go-to accuracy and tracking. If you only use low or medium power eyepieces, then these errors might not be significant, and targets will be placed in the field of view of the eyepiece consistently, but even so, the target won't be in the center of the field. In the case of faint objects like galaxies, or star clusters that don't always stand out from the background field at first glance, it is entirely possible to miss an object not because it wasn't in the field of view, but because it was towards the edge of the field of view and thus overlooked. For lunar and planetary observing where high magnifications are required, accurate tracking is very desirable if you don't want the target to slowly drift out of view, and if you plan on any kind of photography, it is essential.

Therefore, even if you use low magnifications or wide-field eyepieces, it is still worth spending the extra time aligning with high-magnification eyepieces, because this will deliver more accurate tracking and more reliable go-to pointing. What high-magnification eyepieces do is reduce the field of view, so that any errors in your estimate of when the alignment star is in the center of the field become relatively unimportant. Compare a 26-mm Plössl to a 7.5-mm one. When used with a typical 200-mm SCT, a 10 percent error with the former would take an alignment star about a quarter of a degree from the center of the field of view; while the same percentage error with the shorter focal length eyepiece would amount to less than five arc-seconds. This latter is a trivial amount for most observing purposes, but the former is a substantial amount and could easily make the difference between seeing an object and missing it altogether. In addition, when it comes to aligning your telescope, a wide-field eyepiece like a Nagler isn't any better than a narrow field one such as a Kellner. In fact, the narrower the apparent field of view, the easier it is to decide when the star is in the center of the field. The "port hole into space" view provided by wide-field eyepieces demand that you move your eyeball around to take in the entire field, and this makes it very difficult to estimate exactly where the alignment star is relative to the perimeter of the field. So even if you upgrade your eyepieces later on to get one with wider fields or better eye relief, it's worth hanging onto the Kellners or Plössls that came with the telescope, maybe using them with a Barlow lens to ramp up the magnification if need be.

Using a high-magnification eyepiece for alignment is only easy with an accurately aligned finder; otherwise trying to get the alignment star into the narrow field of view can be an exercise in frustration. This is definitely one situation where the zero power finders that come on the NexStar scopes are perhaps less useful than the full sized finderscopes on the LX 90 and LX 200 telescopes. Either way, might want to use a low-power eyepiece to begin with and then swap it for the higher power one for the final stages of putting the star into the center of the field of view. Some observers like to use eyepieces with illuminated reticules to cut out the guesswork altogether. These serve a variety of purposes and come in

Figure 5.5.
Illuminated reticule eyepieces are very useful for aligning go-to telescopes by making it easy to tell when a star is centered in the field of view (photo courtesy of Meade Instruments Corporation).

several different designs, mostly using the Kellner or Plössl design. The key feature to look for is some sort of crosshair that indicates dead center, so using one of these makes it very easy to decide when the selected star is dead center. Though a bit time consuming, a few extra minutes spent getting the alignment stars centered perfectly each time will pay for itself many times over during your observing session.

Don't Weigh Down Your Telescope

This is less of an issue with equatorial mounts than fork mounts because the former have adjustable counterweights to compensate for changes in the mass and distribution of the load, but fork mounts are very sensitive to overloading. Forkmounted telescopes are different because they lack counterweights, and are instead designed to balance the optical tube in front of the pivot with the mirror and eyepieces behind. A friction clutch of some sort engages the motors with the pivot around which the telescope moves up and down; the clutch needs to be disengaged when the tube is moved manually by turning a knob on the outside of the fork arm. This knob can also be tightened to increase the friction in the clutch to compensate for slight imbalances, for example, if a heavier eyepiece than normal is being used that would otherwise cause the optical tube to tip backwards. However, if you put too much stuff on the telescope and so put the telescope severely out of balance, the only way to stop the clutch slipping is to tighten it far more than is safe. Over-tightening of the clutch is damaging because it causes the motors to work harder than they need to, which overheats them out and runs down the batteries faster, and more seriously can cause damage to the gears that make up the go-to mechanism. Even before this happens, go-to accuracy and tracking become noticeably poor as the back end of the telescope slowly sags backwards. This becomes very noticeable if you start adding big 2-inch eyepieces and star diagonals to telescopes such as the NexStar 8i and LX 90 not explicitly designed for astrophotography and with relatively lightly built forks. The larger LX 200 and NexStar GPS telescopes are rather better, and will carry a bit more weight before they start slipping, but still perform much better if counterweights are used instead of over-tightened clutches. Otherwise, avoid using heavy equipment: choose low profile, SCT-friendly star diagonals, for example, and f/6.3 reducer–correctors to maximize the field of view of 1.25-inch eyepieces instead of jumbo-sized 2-inch ones like the big Naglers and Panoptics.

Use Fresh Batteries or an External Power Source

When the batteries on a computerized telescope start running down, the accuracy of the go-to system starts to decline. The same Murphy's Law that holds that buying astronomical equipment always precipitates the longest spell of complete cloud cover for months, so too will the batteries die just when the

Figure 5.6. One of the prime reasons for poor performance by go-to telescopes is insufficient battery power. Although heavy and somewhat cumbersome, external 12V batteries are inexpensive, rechargeable and last a long time even on very cold nights (photo courtesy of Celestron).

seeing has turned good and the things you're interested in are high in sky. So check that the batteries are good before starting. Go-to telescopes work from a set of disposable batteries stored inside the telescope somewhere (usually the base), but the lifespan of these batteries is usually very short, so rechargeable batteries of some sort are an obvious alternative. Sadly, putting rechargeable batteries into the telescope to replace the disposable ones doesn't work very well. In the cold, these batteries frequently supply an inadequate voltage, and even if conditions are favourable, the gradual voltage drop typical of these batteries tends to make go-to progressively less reliable.

An external rechargeable power source using heavy-duty lead–acid batteries works much better and can be a very wise investment. Orion, Celestron, Broadhurst, Clarkson and Fuller, and others are producing "power packs" tailor made for astronomers that not only power the telescope but also come with goodies such as dim red lights for reading maps. You may be able to find one that delivers the right voltage and current for your telescope from a local automotive or hardware store, but check the polarity of the power output jack matches your telescope: at best, if its wrong nothing will happen, but at worst it could fry the circuits inside the telescope. If in doubt, consult your telescope manufacturer or local dealer before taking any chances. Regardless of which power pack you use, the great attraction to these devices is that the big batteries inside them last for hours and are relatively indifferent to low temperatures. They are also very cost effective: costing only two or three times what a set of disposable alkaline batteries would cost. Of course, the downside is that carrying around a heavy power pack as well as the telescope and all your other kit is a nuisance.

Enhancing the Go-To Experience

Optimizing Go-To Performance by Training the Drives

Training the drives of a go-to telescope is very important. All go-to telescopes have a certain amount of slippage and backlash in the motor and gear system, but when trained the computer attempts to compensate for these factors, almost always improving pointing accuracy significantly. First, you center the telescope on a terrestrial object, and then the computer turns the optical tube away from it. After a moment, it then brings the optical tube back again, and then you need to compensate manually for any error. It does this once or twice in both axes of motion (i.e., horizontal and vertical, or right ascension and declination), and once finished the computer has a much better calibration of the relationship between what its motors do and where it is pointing the optical tube. New owners of go-to telescopes anxious to start looking at the night sky at once often overlook this procedure, but really this should be thought of as an essential step.

The exact training procedure will of course vary depending on the go-to system you're using, but normally it works best if the target used is narrow and a reasonable distance away (the sort of thing you'd use to align the finder with the optical tube would be ideal). After all, what you're trying to mimic is the slewing between stars, which are small objects. A telegraph pole away on the horizon works fine. What you don't want is something so large or close that you can't be sure if you center on, and then return to, the exact same point each time the optical tube moves. As with aligning the telescope at night, a high-magnification eyepiece or one with a reticule for determining dead center is very useful indeed. Meade recommend choosing a target that allows you to train have the optical tube pointing about 45° above the horizontal, which places a middling amount of strain on the clutch and motors (there is less strain when the optical tube is vertical, and more when it is horizontal). Naturally, if you routinely use large eyepieces or piggyback cameras into the telescope, or plan to use a counterweight system with a large SCT or Maksutov, fit these onto the telescope *before* training the drives, and leave them on for the rest of the observing session.

An alternative to actually using the view through the eyepiece for training is to employ a laser pointer, such as Howie Glatter's SkyPointer ($165). These fix to the telescope using a bracket ($30 to $40, depending on optical tube). The laser casts a narrow beam of light that extends forwards from the telescope for several hundred meters. Although usable as a finder as well, the key thing here is that it is an obvious indication of which way the telescope is pointing. A marksman's bull's-eye makes a convenient target but any small but obvious feature five meters or so in front of the telescope would do fine. Center the laser beam on the target, and then begin the training procedure. Instead of looking into the eyepiece, look at where the laser beam is, and whether or not it is in the target. Walk up to the target if you like, and draw a pencil mark where the laser beam goes first time, and then use that to check how accurately it returns there on subsequent attempts. Besides training the drive, the laser shows what the telescope is pointing at and a great tool at night if you share observing sessions with others, especially children.

Adding New Objects

Reading over the promotional literature for go-to telescopes in the amateur astronomy press, or the user manuals that come with them, you'd easily imagine that with tens of thousands of objects in the memory banks of the handsets there would be lifetime of observing opportunities built into your Autostar or NexStar telescope. The problem is that the overwhelming bulk of the objects included in these catalogs are stars. Okay, double stars and variable stars can be fun in their way, but after the first few hundred stars do rather all start to look the same. Really these stars are included to bulk up the catalogs and make them more impressive purely as a marketing ploy, and for deep sky observing at least you're left with the NGC and IC catalogs (including subsets of these such as the Messier and Caldwell lists). Admittedly, the NGC and IC are big catalogs, but there are many deep sky objects, particularly galaxies and planetary nebulae, not in those catalogs that big telescopes under dark skies can be used to see. Then there are other objects, like artificial satellites and the International Space Station, that slowly change position over time and the preset positions for them loaded into go-to telescopes at the factory become inaccurate, and anything that wasn't discovered before the telescope was put together, such as new comets, won't be in the catalog either. So for a variety of reasons it can be useful to update the go-to telescope's catalogs or add objects to them to get the most out of the telescope.

Both the NexStar and the Autostar handsets allow the user to add new objects, though the process is fairly tedious when done using the handset as described in their respective manuals. Add new objects one at a time by typing the name and coordinates in using the little buttons on the handset. Much nicer is the ability to create files on a computer or take them from a web site, and then upload them into the telescope handset. Currently only the Autostar system employed by the LX and ETX telescopes from Meade allow this. The Autostar Update Client Application can be downloaded from Meade's web site and run from a Windows PC (or Mac using a Windows emulator). A serial cable capable of connecting the telescope's handset to the computer is also required; Meade sells a USB to RS 232 cable that fits the bill nicely. The Client Application uploads plain text files containing the relevant object information into the telescope handset, allowing much faster incorporation of new objects into the existing database. NexStar users can't easily upload new observing lists to their handsets because the user can't change the memory in the handset. Instead, NexStar users must rely on laptop control, entering the new observing lists into a planetarium program such as *Cartes du Ciel* that will accept additional catalogs, and use the laptop to control their telescope.

Controlling the Telescope Using Planetarium and List-Based Software

As mentioned in Chapter 3, many programs offer some degree of telescope control. All work in basically the same way: the telescope is connected to a laptop computer using a serial cable, and the software on the computer sends signals to the telescope about what object to point at. The planetarium programs tend to be

Figure 5.7. Most go-to telescopes and mounts are controllable from a laptop computer via a serial cable as well as the regular handset. Be sure to get a cable compatible with your telescope: the plug that goes into the telescope varies depending on the model and manufacturer (photo courtesy of Celestron).

point-and-click affairs, where the user sees something interesting in the sky simulation, clicks on it, and the telescope slews to that object. List-based programs work in another way, by allowing the user to create an observing list ahead of time, and then once in the field, allow the user to run through the list with the telescope going from object to object. How useful are these? The answer depends on how frequently you look at objects not included on the Autostar or NexStar lists, or whether or not you have regular sets of targets (such as variable stars or asteroids) that you like to look over at each observing session.

If you simply like to pop outside for an hour or so, then dragging out a laptop and the cables and adapters it needs to connect to the telescope is a hassle. In the dark, wires and cables are trip hazards, and breaking a laptop by pulling it off the bench and onto tarmac is one way to really ruin an evening. Moreover, if all you do is run through the night's best objects as suggested by the telescope's handset or follow some ideas you've found in a book or magazine, then the go-to control offered by a laptop isn't all that useful. In all likelihood you will have a much more productive session simply using the handset alone. If you have a wide-field telescope, like the NexStar 80, then the targets that such an instrument is best suited for are easy to find and well known by their Messier or NGC numbers. Similarly, if you use a small aperture telescope like the ETX 90, there really aren't that many objects worth looking at using these instruments that won't already be in the computer's database (the exception to this is if you have a keen interest in double stars, of which more is said below).

On the other hand, if your observing sessions are more systematic, for example you'd like to concentrate on viewing the Virgo galaxy cluster one night, hopping from one to the next but spending a while at each to take down notes or make detailed observations, then a list-based approach can be very useful. A good resource for planning a theme night like this would be any one of the many deep sky catalogs. I happen to like *The Field Guide to the Deep Sky Objects* by Mike Inglis, but there are lots of others. What makes that particular book so useful to me is that it lists objects visible in any given month, and arranges them from the easier to harder ones, and from their declination you can decide whether or not you'll get a good view from your observing location. This would make it ideal for putting together, for example, an open star cluster observing list for the month of April tailor made for your 200-mm SCT at latitude 53° north. NexStar users wanting to expand their observing beyond the catalogs on the handset are also going to want to use computer control because unlike the Autostar system, the catalogs in the memory banks of the NexStar handsets are not editable.

There are many distinctively colored double and variable stars, such as the silver and gold double 95 Herculis and Y Canes Venaticorum, a blood red variable sometimes called "La Superba", that are either absent from the Autostar or NexStar databases, or included in some less than obvious way. Many of these interesting stars are excellent targets for small aperture telescopes under suburban skies, and there are always dozens of nice ones to look on any night of the year. For whatever reason, the catalogs of double stars that are most popular, such as the Struve and Otto Struve lists, aren't included in the NexStar or Autostar databases. In addition, except for the brightest stars which have their names or Bayer numbers entered, most stars are referred to by the Smithsonian Astronomical Observatory (or SAO) number in the case of the Autostar system, or a proprietary and obscure reference system with the NexStar telescopes. Planetarium programs have all of these stars, and it is very easy to find the star in the sky map, click on it and simply wait until the telescope has slewed across to it.

Maximizing Optical Performance with Collimation

If you own a refractor or a Maksutov, this section covers something you needn't worry about, the adjustment of the alignment of lenses and mirrors within the optical tube, a process called collimation. Only when correctly collimated will SCTs and Newtonians perform well, but when moved about the mirrors and lenses inside these telescopes have a tendency to slip out of collimation. The fact that refractors and Maksutovs don't need this continual adjustment is certainly one reason for their reputation for delivering consistently sharp images. Fast Newtonian telescopes are the most sensitive to poor collimation and need checking at the start of every observing session, while long focal length Newtonians and SCTs can easily hold their collimation for many weeks, even months. Apogee, Celestron, Kendrick, ScopeStuff, and others all produce various collimation aids ranging from simple sighting tubes through to laser collimators and combination Cheshire eyepiece and collimation eyepieces. All can work well, and if you own a Newtonian or an SCT then one or other of these tools is essential. Don't forget though that once you've collimated a telescope it is no longer "looking" in the same direction as it did before, so you will need to adjust the finderscope and go through the alignment process again.

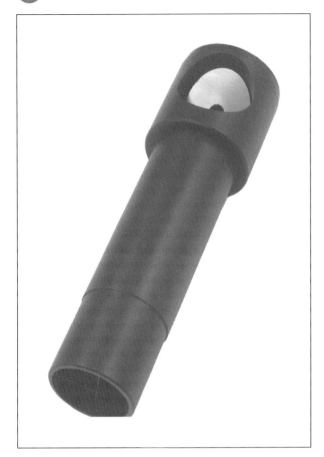

Figure 5.8. A collimation eyepiece is an inexpensive but essential tool for owners of reflector and SCT telescopes. This one combines a Cheshire eyepiece with a sighting tube, but various kinds exist and all of them work well (photo courtesy of Celestron).

Choosing Eyepieces and Other Accessories

Advanced hobbyists often point out that it isn't just the size and quality of the optical tube that matters: the mount and the eyepieces used are equally important. Having looked at the optical tubes and the mounts in turn, the final part of this chapter is about eyepieces. Understanding eyepieces is a complicated business, and the difference between using a good one and a poor one is like night and day. Unfortunately, the best eyepieces are expensive, in some cases approaching the cost of a small go-to telescope package! However, putting aside budget for the moment, there are some crucial points to consider when shopping for eyepieces if you want to get the best performance from your telescope. Are there eyepieces particularly suited to go-to telescopes? Not as such, no, except insofar as high-magnification eyepieces with a narrow field of view, or an illuminated reticule, are useful for aligning the telescope. However, certain optical tubes work better with some eyepieces than others, and if you have a telescope

Table 5.1. It is easy to create a spreadsheet that calculates various parameters for your existing eyepiece collection. This way you know which eyepieces are best suited to certain tasks, and you can identify any gaps that need filling before spending money on new eyepieces. Aperture, focal length, field stop, eye relief and exit pupil are in millimeters; AFOV (apparent field of view) and TFOV (true field of view) are in degrees

	A	B	C	D	E	F	G	H
1	Eyepiece data					Telescope: Meade LX 90 (200 mm SCT)		
2								
3						*f* – number	Aperture	Focal length
4						10	200	2000
5	Eyepiece	Focal length	AFOV	Field stop	Eye relief	Magnification	TFOV	Exit pupil
6						= H4/B6	= C6/F6	=B6/F4
7								
8	Plössl	32	52	27	18	= 2000/32 = 62	= 52/62 =0.8	= 32/10 = 3.2
9	Panoptic	24	64	27	16	= 2000/24 = 83	= 64/83 =0.8	= 24/10 = 2.4

AU: This needs to be reworded as a table title

mounted on a short fork or without a 2-inch focuser, those factors may limit the range of eyepieces still further. Besides mechanical constraints like these, other parameters such as field stop, eye relief and exit pupil are important as well, and not just the more obvious things like magnification and apparent field of view. All these various factors mean that the you will want different eyepieces for viewing the Moon, or looking at globular clusters, or scanning the Milky Way, and that the choices that work on a big NexStar SCT won't necessarily be the best ones for use with Meade ETX 90 Maksutov. So rather than just include a section of the different types of eyepieces and who makes them, this section is focused on determining which eyepieces will work with your telescope, and whether there are specific models that would fit into your observing program particularly well.

One of the best things to do before even looking at the eyepieces at your local astronomy store is to launch your favorite spreadsheet application and create a table that will calculate true fields of view, magnifications, exit pupils and so on based on the characteristics of your equipment. The data on this "eyepiece calculator" spreadsheet will indicate how well a certain eyepiece will fit into your collection. The key bits of information to start with are the aperture and the focal length of your telescope; in the case of a typical SCT like the Meade LX 90, the aperture is 200 mm and the focal length is 2000 mm. From these two pieces of information you can work out the speed, or focal ratio, of the telescope, using the following formula:

$$\text{Focal ratio} = \frac{\text{Focal length}}{\text{Aperture}}$$

It doesn't really matter whether you use millimeters or inches to calculate the focal ratio, so long as you use the same units for both the focal length and the aperture. Don't mix metric units for focal length (e.g., 2000 mm) with Imperial ones for the aperture (e.g., eight inches) or you'll end up with a nonsensical answer! Once you know the aperture, focal length and the focal ratio of a telescope, you can calculate many of the variable parameters needed to understand the performance of a given eyepiece when used with that telescope. You need to know a few of the fixed parameters first though, the focal length and the apparent field of view being most important. To obtain the magnification delivered by a given eyepiece when used with a certain telescope use the formula that follows, again being sure to use millimeters for both parts of the fraction:

$$\text{Magnification} = \frac{\text{Focal length of telescope}}{\text{Focal length of eyepiece}}$$

The true field of view (i.e., how much of the sky you see through the eyepiece, in degrees) by dividing the apparent field of view, in degrees, by the magnification as calculated previously:

$$\text{True field of view} = \frac{\text{Apparent field of view}}{\text{Magnification}}$$

Another important relationship is between the eyepiece focal length and the telescope's focal ratio. This is the exit pupil, calculated using this formula:

$$\text{Exit Pupil} = \frac{\text{Eyepiece focal length}}{\text{Telescope focal ratio}}$$

The exit pupil is the size of the image formed by the eyepiece: the longer the focal length, the bigger the exit pupil. Generally speaking you don't want an exit pupil any larger than the widest the pupils of your eye can expand to, which in a young person is around 7 mm, though it decreases by a millimeter or two as we age. An exit pupil above this limit casts a bigger image than the eye can take in, which in theory wastes the light that doesn't enter the eye and stimulate the retina. In addition, an oversized exit pupil allows the secondary mirror of reflecting telescopes (including SCTs and other catadioptrics) to become apparent too. But if you have a refracting telescope the "wasted light" argument can be ignored because the resulting low power, wide field view is so dramatic and impressive in its own right and there's no secondary mirror shadow to get in the way of the image. This is one reason why rich-field refractors are so popular: they are the only one of the popular telescope designs that work well with the 5 to 7-mm exit pupils produced by using low power eyepieces like the 55-mm Plössl or 31-mm Nagler. On the other hand, a 5-mm or larger exit pupil used under suburban skies rather than dark ones can be a bit disappointing, regardless of the design of telescope used. Under these circumstances, the background sky glow becomes a distracting part of the image because it weakens the contrast between the thing you're looking at and the sky. If there is mild light pollution, decreasing the exit pupil to 3 to 4 mm will have the effect of darkening the sky and improving contrast a bit, though at the cost of an overall dimming of the image. A rich field refractor such as the NexStar 80 GT delivers a fine 3.2° swath of the sky with a 26-mm Plössl, but as the exit pupil is 5.2 mm, this combination will be sensitive to ambient light pollution. Swapping that Plössl for a 19-mm Panoptic gets you the same true field of view, but with an exit pupil of only 3.8 mm, and consequently darker skies and better views when used somewhere with mild light pollution. However, while this trick works well with open clusters and star clouds, it isn't so helpful with extended objects like nebulae and galaxies, which become too faint to observe clearly. For those objects, there really is no substitute for good, dark skies (although a nebular or light pollution filter may help).

Globular clusters, planetary nebulae and galaxies respond well to small exit pupils, many astronomers considering an exit pupil of 2 mm being ideal. As far as resolving details go, the eye performs best when the exit pupil is around 2 mm, since this results in an image on the retina that fully occupies the region composed solely of cone cells, which as discussed in Chapter 3 are the ones we use to see details. So although a smaller exit pupil may cast a larger image on the retina, it will cross not just cone cells but rods too, and so seeing the details will be more difficult; this is where part of the "learning to see" mantra recited by many experienced observers becomes relevant. The brighter globular clusters, for example, look particularly good under suburban skies with an exit pupil of 1.3 to 1.5. A 12.5-mm Plössl would deliver an exit pupil of this size when used with a 200-mm f/10 SCT. Very small exit pupils, less than 1 mm in diameter, come from using

Figure 5.9. Plössls are the most popular moderately priced eyepieces, and on telescopes slower than f/5 they generally work very well. Their main problem is limited eye relief at focal lengths less than 10 to 12 mm (photo courtesy of Meade Instruments Corporation).

short focal length eyepieces. Though these deliver high-magnification images suited to lunar, planetary and double star observing, on most nights they are of limited use because the resulting tiny images are extremely difficult to focus. Unless the sky is very steady, the image will seem to blur in and out of focus almost as if it was boiling, in which case it is best to revert to a lower magnification with an exit pupil between 2 to 4 mm. But if the skies allow them, high magnifications combined with small exit pupils can, in theory at least, reveal as much detail as the optics of a telescope will deliver, which depends as much on the atmospheric conditions as the optical system. Since deterioration of the image due to seeing conditions increases with aperture size, a telescope that makes the most efficient use of a modest aperture will be the design best suited to high-magnification observing on those nights that don't have perfect seeing, and that design is the apochromatic refractor. This is one reason why those telescopes are so favoured; they don't magically break any rules of physics relating resolving power to aperture, it's just that on the average night they will deliver sharp, high-magnification images when the view through other, larger telescopes is only mediocre even at medium powers.

Two other eyepiece criteria are important as well, the field stop and the eye relief. The field stop is the diameter of a metal ring inside the eyepiece that sets a limit on the size of the image. Although in general it gets bigger as eyepiece focal length increases, it also varies somewhat with design: an 8 mm Plössl has a field stop of 6.5 mm, whereas an 8 mm Radian has a field stop of 8.3 mm. The maximum field stop possible with the 1.25-inch fitting is 27 mm, which Plössl eyepieces reach at the 32 mm focal length mark. There will show you as wide a field of view as is possible with a 1.25-inch eyepiece, for example about 0.8° when used with a 200 mm f/10 SCT. Of course, one way to improve the image is to raise the magnification while keeping the same true field of view; this would improve the contrast of deep sky objects in particular, as well as deliver a more dramatic, immersive view. The 24 mm Panoptic would do exactly this: with the 200 mm SCT you would still get the 0.8° field, but at ×83 rather than ×63. Depending on your budget, either of these eyepieces would be a very sensible upgrade to a telescope that comes with a 1.25-inch fitting. To get an even wider field of view, you need bigger field stops, and these are only possible with 2-inch eyepieces. Not all telescopes will accept these (the NexStar 114 and ETX 90 won't, for example) but many will, either out of the box, as with the LXD 55 Schmidt–Newtonians or by upgrading the visual back to the 2-inch fitting, as is usual with 200 mm SCTs. When picking out low powers, you want ones with a field stop around 1.5 to 2 times higher than the one preceding it. So if you have a 20 mm Plössl (with a field stop of 17.1 mm), a field stop twice that, i.e., about 34 mm, might be appropriate. An eyepiece that would fit this bill is the 26 mm Nagler, with a field stop of 35 mm. Alternatively, you might prefer two lower power eyepieces after the 20 mm Plössl, separated by a ratio of 1.5 instead, in which case they would need field stops of about 26 and 38 respectively. Among the possible eyepieces that would work are the 32 mm Plössl mentioned earlier (with a field stop of 27 mm) and the 35 mm Panoptic (which has a 38.7 mm field stop).

All these numbers and equations may seem baffling at first glance, but once they are down on a spreadsheet, they can go a long way towards streamlining

Figure 5.10. Premium eyepieces generally offer wider, flatter fields and better color correction in fast telescopes than tradition designs like Plössls and Kellners; many also feature much more generous eye relief at short focal lengths as well (photo courtesy of Tele Vue Optics, Suffern, NJ).

your eyepiece collection. By looking at magnifications, exit pupils and fields of view you can determine which eyepieces in your collection will be most useful for use on whatever your observing targets are on a particular night, and any obvious gaps in the roster will be helpful when the time comes to buy new eyepieces. The first thing to look at is the range of magnifications: except on rare nights of perfect seeing, powers above 200 times are not useful, but a range of eyepieces between ×25 to ×200 will be useful for almost any telescope. Exit pupils should be next on the list of things to check out, as indicated before there isn't any point using an exit pupil above 7 mm with a reflecting telescope, and unless you regularly have good seeing at your observing locality, you won't need very many eyepieces that deliver exit pupils of less than 2 mm. One way to expand the versatility of a telescope is not to add eyepieces but to use Barlow lenses and focal length reducer–correctors instead. These lenses change the way the light passes out of the telescope before it hits the eyepiece, in the case of a Barlow lens increasing the effective focal length of the instrument, and with a reducer–corrector decreasing it (and flattening the field to boot). A typical Barlow lens doubles the focal length, turning an f/10 SCT into an f/20 one. So if a 26 mm Plössl delivers a magnification of ×77 with a 200 mm SCT, the magnification goes up to ×154 when the Barlow lens is used. Exit pupil changes, too, from 2.6 mm, which is nice for galaxies, down to 1.3, an ideal size for observing globular clusters. The reducer corrector does the opposite, typically turning an f/10 SCT into an f/6.3 one instead. The same 26 mm Plössl will now produce an image with a magnification of ×49 and an

Figure 5.11.
Reducer–correctors are great accessories for owners of SCTs, flattening the field and increasing the field of view considerably (photo courtesy of Celestron).

exit pupil of 4.1 mm, a nice size for use under mild light pollution. What is so great about both of these extras is that neither is expensive (good examples of each being around $100). Note, however, that reducer–correctors are for use 200 mm and larger SCTs, only (they may work with other catadioptrics too, but check with your dealer to be sure).

Barlow lenses are great all-round purchases, and most amateurs have at least one in their collection regardless of the type of telescope they use. Besides being good for increasing the magnification and shrinking the exit pupil, it is generally nicer to use than a medium focal length eyepiece with a Barlow than a short focal length eyepiece alone. Short focal length eyepieces, unless especially designed otherwise, have limited eye relief, meaning that to see the full image the eye needs to be close to the glass. Spectacle wearers generally need at least 15 mm of eye relief to find an eyepiece usable at all, while those without can get by with much less. For non-spectacle wearers, eye relief as low as 10 mm can be perfectly comfortable, but below that, things start to get unpleasant very quickly. Plössls, Kellners and orthoscopics eyepieces typically have uncomfortably short eye relief below a focal length of around 12–15 mm, depending on the design; while the Radians and Vixen Lanthanum eyepieces maintain a fixed eye relief of 20 mm regardless of focal length. Short focal length versions of the fancy, wide-field models, such as Celestron Axioms, Naglers and Meade Ultra Wide Angles, come somewhere in between, and are usually fine for people who don't wear eyeglasses but sometimes awkward for those who do. It is certainly well worth including a column in your eyepiece spreadsheet for eye relief values taken from the manufacturers promotional literature or your dealer, and then testing out the eyepieces in the field to see what eye relief values are most comfortable for you.

Kellner, Orthoscopic and Plössl Eyepieces

Kellners are the lowest priced eyepieces of any real usefulness to the amateur astronomer; there are other kinds such as Huygens that are seen from time to time, but these have so many shortcomings that they are best ignored. Briefly, Kellners work well with slow telescopes such as SCTs where they can deliver sharp images (at least in the center of the field of view) without the false color that characterizes the more primitive eyepieces. Their main problem is a tendency towards ghosting, internal reflections that create a haze around bright objects such as planets, particularly with large aperture instruments. How annoying this is depends on the observer, some people get used to it, while others find it very intrusive. Other problems are a narrow field of view, for example, 45° in the case of the Celestron Kellners (called "Super Modified Achromats"), and the painfully short eye relief of eyepieces with 10 mm and lower focal length. The modest aperture, long focal length and automatic tracking of the 125 mm and smaller go-to Maksutovs and SCTs in particular would largely mitigate the various weakness of the Kellner design, making them very viable choices for budget-conscious owners of those telescopes. Kellner eyepieces are usually only included with entry-level telescopes, such as the Meade ETX 70. Good orthoscopics have a similar field of view to Kellners, but better contrast, sharpness, color fidelity and much less ghosting. Eye relief at short focal lengths is a little better, too. Again, they work best with telescopes of long focal length, but so used they are among the best planetary eyepieces there are, and surprisingly among the least expensive designs as well. Plössls are the eyepiece of choice for most amateurs: they are inexpensive, exhibit little ghosting and deliver sharp, contrasty images in slow and medium speed telescopes. The best will even work in quite fast, f/5 telescopes. Most have a field of view slightly larger than a Kellner or orthoscopic, around 50°. Plössls range in price from $30 to $250 depending on the design and focal length. Though the cheap ones can be real bargains, it is worth spending a little more money to make sure the eyepiece has fully multi-coated optics. Other niceties include an eyecup to keep out stray light and rubberized barrels to make them easier to handle when it is cold and dark. Most Celestron and Meade go-to telescopes come with at least one Plössl eyepiece, the Meade ones being in fact modified Plössls, and noted for their particularly sharp and flat images.

Wide-Field and Long Eye Relief Eyepieces

Though many amateurs are quite happy to build up a set of Plössls, orthoscopics or Kellners, others prefer eyepieces that offer wider fields of view and longer eye relief. As noted earlier on, higher magnification has the effect of darkening the background sky, so using a higher power but wider field eyepiece than normal can be a good way to mitigate the effects of moderate light pollution. Replacing a 32-mm Plössl with a 24-mm Panoptic will do exactly this, these two eyepieces showing the same amount of sky but with the Panoptic having a broader but more magnified view than the Plössl. I also mentioned that another reason for using these deluxe eyepieces is the better eye relief compared with the traditional designs, something that can be critical with short focal length eyepieces. But for many amateur astronomers the key thing is that these eyepieces are "highly corrected", meaning that the problems that can occur in the older designs such as ghosting, false color, curvature to the field of view, and an overall lack of sharpness, are much diminished, usually absent, from these eyepieces – even in very fast telescopes! For many people these are the best eyepieces for observing of all types and with any telescope.

The problem is that once you start using these deluxe eyepieces there's no going back, and your non-wide-field eyepieces end up being retired or sold off. What's using a wide-field eyepiece like? Imagine comparing the view through a keyhole and then through a porthole. Using a narrow field of view eyepiece, like a Kellner, is like looking through a keyhole, the view is obviously limited, and you can only see things straight ahead of you. A wide-field eyepiece is like the porthole, there is still a boundary to what you can see, but it is much bigger. Advocates of wide-field eyepieces call this the "spacewalk" impression. I'm not sure I'd go quite this far, not having walked in space, but certainly the view of a star cluster through a big wide-field eyepiece is something like how I imagine the view through the window on a spaceship to be. Views of the Moon can be wonderful too for the same reason, but on the planets or double stars these eyepieces are largely redundant, where an equatorial mount or computerized tracking easily keeps small objects in the field of view for as long as you want. Celestron, Meade, Pentax, Tele Vue and Vixen all produce lines of wide-field eyepieces, ranging in price from $150 to $600. Besides cost, there are a couple of other factors worth considering. Firstly, as noted earlier on, the widest fields only come with 2-inch barrels, and not all telescopes can accept these. They can also be very heavy, some exceeding 0.5 kg (1 lb) in weight, which can be enough to mess up the tracking and go-to accuracy of the lighter weight fork-mounted telescopes such as the LX 90. Any of these deluxe eyepieces is an investment for a lifetime's enjoyment of the hobby, regardless of how your observing interests change and you expand your collection of telescopes, these eyepieces will work consistently well. However, at several hundred dollars a pop, it is worth taking good care of them by storing them safely in an eyepiece box of some sort. Hardware stores and photography shops sell aluminum equipment cases and flight cases that work very well. If the worse happens and you damage an eyepiece, you may be able to get it repaired, depending on the company. For example, Tele Vue does repairs, while Meade does not.

Besides wide-field eyepieces, Tele Vue and Vixen also produce long eye relief eyepieces, the Radian and the Lanthanum series respectively. Both of these offer 20 mm of eye relief, which is plenty even for people who wear spectacles when observing, right down to amazingly short focal lengths such as the 2.5 mm Vixen ($120). Planetary observers like the high-power views Lanthanums give, but the image is the tiniest bit dimmer than through a traditional eyepiece of the same focal length. The field of view, 50°, is also rather small compared to most of the other deluxe eyepieces. If you want something as bright and sharp as the best orthoscopics, and with a wider field, then the Radian eyepieces ($240) with their 60° fields of view are worth considering, having a reputation for being among the very sharpest eyepieces around. Like the Lanthanums, they all come in the 1.25-inch fitting, so they work well with big and small telescopes. For a planetary or double star observer wanting crisp, sharp views with comfortable eye relief these are a very good choice, though expensive. Fortunately, Vixen offer a slightly cheaper alternative to the Radians in the form of their Superwide Lanthanums, which offer a sharper, brighter view than the original Lanthanums, and bigger fields too; at 65° being comparable to a Panoptic or Meade Super Wide.

CHAPTER SIX

Webcam and Digital Camera Astrophotography

Webcam astrophotography is fast becoming the standard way for newcomers to the hobby to break into astrophotography. The traditional method, using 35-mm SLR cameras and regular film, is fiddly, time-consuming and above all wasteful. It takes a lot of practice to get consistently good pictures, and even then, many exposures simply aren't worth keeping. CCD, or *charge coupled device*, cameras have for the last few years become steadily more widely used and have the advantage of being digital, cutting out the wastage of film typical of traditional astrophotography. However, CCDs are very expensive, even the simplest ones costing in excess of a thousand dollars and more versatile, higher performance ones, such as color and wide-field ones, costing several times that. As a result, CCD cameras remain very much the playthings of dedicated amateurs (with understanding bank managers or shares in a platinum mine!). Webcams in contrast are very inexpensive, basic ones costing about as much as a generic Plössl eyepiece, say around fifty dollars, and even the best ones not much more than that. Webcams are also widely available and compatible with Windows, Mac and Linux computers, even relatively old ones, since in general webcams don't demand much in the way of processing power. Another great advantage of webcams is the sorts of images they do well, namely shots of the Moon and brighter planets, are exactly the ones that budding astrophotographers are going to want to image using relatively modest telescopes and mountings. For all these reasons, webcams make a logical first step for amateurs looking to capture something of the excitement we feel when viewing the rings of Saturn or seeing sunrise creep along the lunar terminator. More ambitious photographers can make animations by grouping successive stills into animated GIFs or QuickTime movies, perhaps showing the rotation of Mars or the transit of a Galilean moon across the face of Jupiter. Besides the Moon and planets, webcams are often sensitive enough to image bright stars. Double stars with their contrasting colors make a very nice subject for a digital image catalog: finally, a way to decide if those stars are topaz and garnet, or orpiment and chalcedony!

Choosing Webcams for Astrophotography

CMOS Versus CCD

Many people think that webcams contain low-cost CCDs, but strictly speaking, only a few actually do. Webcams more frequently contain a *complementary metal oxide semiconductor*, or CMOS, chip instead. The essential difference between the two is how they send the charge they produce when exposed to light to the circuit board they're mounted on, the CCD being an analogue device is this regard and the CMOS a digital one. When light hits a CCD chip, the pixels send a charge as an analogue signal to the rest of the circuit board that converts into the digital signal the computer uses. A CMOS chip makes this conversion from charge to signal on the chip itself, and sends out a digital signal that's ready to use. This makes a CMOS chip a much more integrated unit and consequently webcams built around them can be smaller. The downside is that some of the surface area of the CMOS chip has to include the hardware that makes the charge to signal conversion. This inefficiency makes it less sensitive compared to a CCD chip, and reduces the quality of the images it produces. When comparing webcams with CCD and CMOS chips of the same surface area inside them, the CCD one can be expected to produce better images and work more effectively under the low light conditions typical of astrophotography. The Creative Labs Webcam Pro, the Logitech QuickCam Pro and the Orange Micro iBot for example are CMOS

Figure 6.1. Out of the box, webcams such as the Logitech QuickCam 4000 Pro can be used to take nice pictures of the Moon and brighter planets at a fraction of the cost of dedicated astrophotography CCD cameras (photo courtesy of Logitech).

webcams, while the Unibrain Fire-i, the Logitech QuickCam Pro 3000 and the Philips ToUCam Pro are all CCD cameras. To keep things simple, I'm going to use CCD throughout this chapter to refer to the light sensitive chip inside the webcam, but it is just as well to know the differences between them and use that information when shopping for a webcam to put to astronomical use.

CCD Chip Hardware Resolution

Just as important as the right sort of webcam with a sufficiently sensitive chip inside it, is getting a webcam with a decent hardware resolution. This isn't the same thing as software resolution, which is the resolution of the image as you choose to record or display it. The actual size of the chip (some are advertised as being "quarter-inch" or whatever) is unimportant: what is important is how many pixels, counted by row and column, there are on the actual chip. The more pixels, the higher the hardware resolution, and the better the images will be since you can now record more detail per frame. Most modern webcams will offer 640 by 480 pixel resolution, sometimes called VGA resolution, and this is ideal for most purposes. Some have less, for example, the Creative Labs Webcam has only a 352 by 288 hardware resolution, and such a webcam would be poor choice for astrophotography. Webcams sometimes offer *software interpolation* for multiplying up an image from a low resolution to a higher one. This isn't much good either for astrophotography even if it might work adequately well for videoconferencing and daytime photography. The software simply doesn't do a good job of "creating" detail on small, faint objects like planets that occupy just a tiny portion of the CCD. It is much better to stick with cameras with hardware resolutions of at least 640 by 480 so that software interpolation won't be necessary.

Webcam Interfaces: USB, FireWire and the Rest

The interface is the connection between a peripheral, such as a webcam, and the host computer. There are many such interfaces, but the main distinction between them is how much data can pass along them between the peripheral and the computer. This is the data transfer rate and is measure in megabits per second. All modern webcams use either the *Universal Serial Bus* (or USB) interface or *FireWire* (also known as *i.link*, and less poetically perhaps, as *IEE 1394*). Computers at all price points and running all operating systems generally come with the USB interface, whereas FireWire is standard only on the Macintosh and high-end Windows PCs or as an upgrade PC card. Besides webcams, USB is used for devices which do not need high data transfer rates, such as mice and keyboards, whereas FireWire tends to be used for things like digital camcorders and high-performance external drives, where large amounts of data need to be transferred quickly.

There are two types of USB, referred to as USB 1.1 and USB 2. USB 1.1 is relatively slow, with a maximum data transfer rate of about 12 megabits per second but this is the variety most commonly seen in home computers. This data transfer rate

Table 6.1. Different types of webcam based around the various interfaces used by modern home computers. Of the current designs, USB 2 and FireWire offer the best performance

	Supported hardware	Hot pluggable?	Data transfer rate	Uncompressed full color 640 by 480 video possible?	Price
ADB	Older Macintosh	No	< 1.6 Kb/s	No	Obsolete
Serial	Windows & Linux	No	< 15 Kb/s	No	< $50
Parallel	Windows & Linux	No	< 100 Kb/s	No	< $50
USB 1.1	Windows, Macintosh & Linux	Yes	~ 12 Mb/s	No	$50–100
USB 2	Windows, Macintosh & Linux	Yes	~ 480 Mb/s	Yes	$100 +
FireWire	Windows, Macintosh & Linux	Yes	400 or 800 Mb/s	Yes	$100 +

is a bit more than floppy disk's worth per second. USB 2 is much faster, with a theoretical data transfer rate of up to 480 megabits per second, or about as much as forty or so floppy disks. The majority of budget webcams use the USB 1.1 interface, with USB 2 reserved for the more expensive ones, particularly those designed for videoconferencing. FireWire also comes in two flavors, known as FireWire 400 (IEEE 1394) and FireWire 800 (IEEE 1394b). The numbers refer to the speeds of data transfer, up to 400 and 800 megabits per second respectively. Only high-end webcams use FireWire 400, which has a transfer rate about the same as 35 floppy disks per second. It is therefore far superior to USB 1.1 webcams but comparable to those relying on USB 2. FireWire 800 is twice as fast as FireWire 400 and so promises very high data transfer rates, but so far hasn't appeared on any consumer-grade webcams (as of the time of writing at least) but may do in the near future.

What makes the data transfer rate significant is that with USB 1.1, the camera must compress the images first and only then send them down the cable to the computer. This keeps the amount of data carried within the capacity of the interface, but because compression is lossy (meaning that some data must be thrown away to make the image smaller), the quality of the image decreases. When imaging planets in particular, the object only covers such a small part of any given frame that any loss of detail is significant, so you want to keep compression down to a minimum. Normally, USB 2 and FireWire webcams don't compress the image at all, meaning a much higher quality movie gets recorded by the computer, making the job of turning movies into high-detail pictures much more easy. All else being equal then, a FireWire or USB 2 webcam makes a better choice for astrophotography than a USB 1.1 webcam.

Older webcams used different interfaces, such as the parallel port on Windows PCs or the Apple desktop bus (ADB) interface on older Macintosh computers. If you have webcams such as these they are certainly worth experimenting with, but none match the performance of even the USB 1.1 webcams, so inevitably compression becomes even more of an issue and the quality of what you will record suffers accordingly. The other great advantage of USB and FireWire webcams is that they are *hot pluggable*, meaning they can be plugged and unplugged without the computer needing to be turned off.

Frame Rates, Resolution and Exposure Times

Virtually all webcams advertise their frame rates on the packaging. A frame rate of thirty frames per second is what needed for that motion picture feel, and the more good frames you can grab in those fleeting moments of good seeing, the better your final image is going to be. On the other hand, the USB 1.1 interface simply can't handle uncompressed true color, high-resolution video (i.e., a movie made up of 640 by 480 pixel frames in millions of colors). A single uncompressed, or *raw*, frame of this quality is about 1.2 MB in size, or 9.6 megabits, so to transfer thirty of them a second the cable between the camera and computer would need a capacity of at least 288 megabits per second. This is well over the 12 megabits per second limit of USB 1.1, and to get this quart into a pint pot is where the compression mentioned before comes into play. USB 2 and FireWire can both carry well over 288 megabits per second, so cameras with these interfaces don't need to compress the images before sending them down the cable. Therefore, simply relying on the frame rates as quoted on the box, without understanding the limits the interface sets on image quality, can be very misleading. A second important benefit of a high data transfer rate is that focusing becomes easier. Focusing the telescope with a webcam in place is a bit tricky because you need to go by what you see on the screen, which isn't particularly intuitive to begin with, and factor in a lag between what you do with the telescope and what appears on the monitor and it becomes even more difficult.

One way around the bottleneck in the rate at which data moves from the webcam to the computer is to reduce the resolution of the images. A movie with a resolution of 320 by 240 pixels will require only one quarter the data transfer rate as one 640 by 480 pixels in size. The downside to this is that image will be spread out over fewer pixels, making the image blockier and less attractive. This is the same effect as looking at your television screen from a distance and then close up: from the couch the images look fine, but get up close and the individual blobs of color become obvious. Reducing color depth, or *bit depth*, can be another way to improve the performance of USB 1.1 cameras. Black and white imaging is fine for the Moon, where there isn't much color anyway, and uses far less capacity than full color. With the planets and stars, where color is important, you can often get away with turning down the bit depth from true color (32+ bit) down to thousands of colors (16 bit), which will improve the rate of data transfer without losing much detail.

Many webcams come with software that allow you to change the exposure length. On a regular SLR camera, you do this by changing the length of time you leave the shutter open for, but webcams do not have a shutter, and instead "exposure length" refers to how long the CCD chip is exposed to light before it sends a batch of information to the computer. The longer the exposure, the more time the CCD has to absorb light and convert it into signal, and the better it can image a faint object. Typically, webcams can offer exposures no longer than one-thirtieth of a second, much shorter than the many minutes used by high-end astronomical CCD cameras and 35-mm film cameras. A small fraction of a second simply isn't long enough to be of much use for imaging deep sky objects, and instead webcam users must stack a series of short exposures to mimic a single longer exposure, a process discussed in more detail later on. Incidentally, *frame rates* are *not* the same thing as exposure lengths, and so cutting down the frame rate to 4 frames per second doesn't mean each frame will be exposed for a quarter of second; rather, the it means there will be quarter second delay between each exposure, which might only last a thousandth of a second.

Eyepiece Projection Versus Prime Focus Photography

Eyepiece projection photography, as its name suggests, captures the sorts of image cast by an eyepiece that we see with our eyes. This is the usual approach taken by amateurs using digital still cameras (discussed at the end of the chapter). This is great if you have a low-power eyepiece that frames the entire surface of the Moon nicely, and you want a single shot to record that. 26 to 32-mm Plössls are particularly popular for this sort of photography, and specially made ones with adapters for SLR and digital cameras are available; listings of distributors of these are in Appendix 1. Changing eyepieces allows the user to increase or decrease the

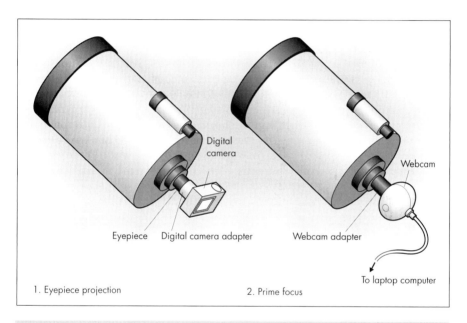

Figure 6.2. In eyepiece projection photography, the telescope is set up normally, but when you have the image framed nicely in the field of view of the eyepiece, you replace your eye with a digital still camera by either holding it very steadily over the eyepiece lens or using some sort of adapter to secure it in place. Prime focus photography is different, with the eyepiece removed after the image is framed correctly and the webcam locked into the eyepiece holder via a webcam adapter. This latter approach is fiddly, but versatile.

magnification as required. In use, focusing is relatively easy because the overall magnification is low and the camera or webcam comes into focus close to the point where it would be in focus for visual observing. This makes the eyepiece projection technique compatible with all telescopes.

In contrast, *Prime focus photography* places the webcam at the point where the telescope forms the image before the eyepiece magnifies it, the *focal plane* of the telescope. This is often referred to as the "prime focus" of the telescope, and hence the name. Instead of being dependent on the magnification of an eyepiece, the size of the image at prime focus depends on the focal length of the telescope, adjusted using reducer/correctors or Barlow lenses as required. Focusing is much trickier than with eyepiece projection photography because the webcam comes into focus at a point well away from the normal visual observing focal point. Indeed, it may be impossible to focus a webcam used this way with some telescopes (such as short focal length refractors). Inserting a Barlow lens between the webcam and the eyepiece holder *usually* helps by moving the focal plane of the telescope, but with it comes increased magnification which exacerbates the usual problems from bad seeing, ambient vibrations and so on. A highly magnified planet will also be fainter than when viewed at low magnification, though it is unusual for the for the image to become so faint the webcam ceases to pick it up; on the other hand, this problem is serious when you begin to image things like double or nicely colored stars.

Both prime focus and eyepiece projection photography have their strengths: the low-magnifications of eyepiece projection being best suited to wide-field deep sky and shots of entire lunar or solar disc, while prime focus photography is more suitable for zooming in on details on the Sun or Moon or for detailed views of the planets. Most of this chapter is about prime focus photography using webcams; discussion of eyepiece projection is towards the end in the section on digital cameras. You *can* use a webcam in eyepiece projection mode though by cementing the inside ring of a T4 camera adapter onto the front of the webcam. The webcam will now fit onto the 35-mm camera adapters widely sold for eyepiece projection use.

Modifying Webcams

Although this chapter is about unmodified webcams, i.e., webcams used pretty much as they come in the box, webcam modifications are quite popular among those who get into webcam astrophotography seriously. However, making them isn't easy by any means, and the exact methods vary with each design of webcam. The aim is to increase the exposure lengths possible, usually by removing or changing some component on the webcam's circuit board, and typically adding a cooling device to improve the quality of the images. This latter is necessary because as the exposure length increases, so does the electrical noise recorded by the CCD and turned into blurs and crackles visible on the image. With cooling this noise can be minimized, and the results of modified webcams can be most impressive, allowing users to cast their nets further than simply the Moon and planets and onto globular clusters, nebulae and so on.

Making a modified long exposure webcam isn't for the faint hearted or those who lack experience of taking apart and soldering together small electronic components. It is very easy to destroy a webcam if you don't know exactly what you're doing, and it renders the webcam unsuitable for normal use. A number of web sites describe these modifications, and Appendix 1 lists some of the best. An alternative to making your own is to buy a professionally made modified webcam: SAC Imaging offer a range of CCD cameras that based around consumer-grade webcam CCD chips, priced from around $250 upward. They are more versatile than regular webcams, the more expensive ones including color imagining, better resolution and cooling systems for less noisy images and longer exposures. On the other hand, they don't compete with the astronomical CCD cameras sold by Meade, SBIG, and others that are more sensitive to light and produce images that are much more detailed.

Webcam, Laptop and Telescope: Getting Them Working Together

Before a webcam will work with a computer, you will need to install the correct *driver software* from the CD-ROM that came with webcam. Driver software expands the system software, i.e., Windows XP or whatever, so that it recognizes the webcam hardware and care record movies through it. Installing drivers isn't usually very difficult: normally a matter of inserting the appropriate disk, running the installer program and following the onscreen instructions. However, it is always well worth checking if the manufacturer has released newer versions of the drivers since the CD was recorded: quite often problems with computer hardware go away when updated drivers are installed; conflicts between outdated hardware drivers and newer versions of the operating system software are notoriously common and troublesome to fix. The next step is to install the movie recording software. Virtually all webcams come with Windows drivers and movie software, but only a subset of these come with Mac software as well, in which case third party software will be needed. Linux users enjoy little support from commercial webcam manufacturers and will almost certainly need to find both drivers and movie recording software from third party sources.

With the software installed, test out the webcam to check it works properly. It is much easier to troubleshoot the webcam during the day and when you have an Internet connection to download any additional software fixes than it is by night under the stars. Assuming you can record and view movies using the webcam without any problems, the next step is to connect it to the telescope to begin using it astronomically. The webcam (with its lens removed) plugs into the eyepiece holder instead of an eyepiece. Removing the lens from the webcam is therefore the first step to getting your imaging set-up together. In some cases, as with the Philips ToUCam (Figure 6.3), the lens screws out easily and is replaced with an

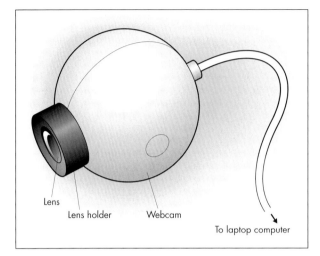

Figure 6.3. Most webcams have the same basic design, a plastic case containing the CCD or CMOS chip, with a lens at the front and the USB or FireWire cable at the back. For prime focus photography, the lens needs to be removed and replaced with a webcam adapter.

Lens

Lens holder   Webcam

To laptop computer

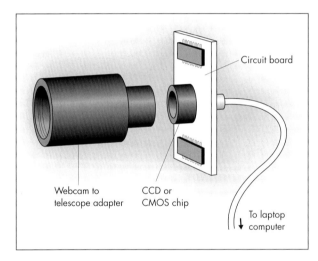

Circuit board

Webcam to   CCD or
telescope adapter   CMOS chip

To laptop
computer

Figure 6.4. In some cases the webcam lens screws out easily, but in other cases the webcam will need to be disassembled first. Either way, this exposes the CCD or CMOS chip to dust and moisture, so proceed carefully. The webcam adapter usually screws or clips into place directly over the chip; be sure and choose the right adapter for your webcam.

adapter, but with others, such as the Logitech webcams, the entire webcam must be disassembled (Figure 6.4). Other webcams are taken apart in different ways. Appendix 1 includes some links to astrophotography web sites that give advice on adapting webcams.

With the lens removed, an adapter connects the webcam to the eyepiece holder of the telescope. An excellent way to do this is to use the webcam adapters, such as those made by Steven Mogg (see Appendix 1). Made from black plastic, these screw into the lens holder of the webcam at one end, and at the other form an eyepiece-shaped tube that the telescope eyepiece holder grips firmly.

Before putting the webcam into the eyepiece holder, you will need to center the target in the field of view. Use an eyepiece of moderately high power (say, a magnification of a hundred or so) to align your telescope to the object. You want to place the target dead center, and just like aligning a go-to telescope, the higher the magnification you use, the more accurate your results will be. On a 2000-mm SCT, for example, this step is a lot more productive using a 20-mm Plössl than a 32-mm one. If you have a reticule eyepiece, this is a good time to use it. A properly aligned finderscope is useful as well, so that you can check the telescope points squarely at the target even with the webcam plugged in instead of an eyepiece. There is an alternative to aiming the telescope visually and then removing the eyepiece and replacing it with the webcam, and that is to use a *flip-mirror system*. Most of the telescope manufacturers make these and cost around $150 to $300 depending on whether they are in the 1.25-inch or 2-inch format, and allow both an eyepiece and a camera or webcam to plug into the eyepiece holder. By toggling the flip-mirror, the light goes to *either* to the eyepiece *or* the camera and this makes it very easy to aim the webcam properly.

Once you are satisfied with the alignment of the telescope with your target, swap the telescope eyepiece with the webcam and launch the appropriate image capture software. There's no need to *record* any movies just yet, all you need to do is make sure you have a clear image on the screen. Most likely, all you will see is a white or grey blur; this is because the focus point for eyepieces is very different to that of a webcam. This is where making sure the target was dead center in the field of view becomes important. Since you probably won't be able to see the target on the laptop screen at first, as you focus one way or another it is very largely your accuracy in aligning the system that will determine whether or not the Moon, planets or whatever come into view. For this reason it is best to practice using the Moon rather than one of the planets, since even when out of focus it is big enough and bright enough to make its presence known. Jupiter isn't a bad alternative, and though much smaller it is quite bright and when out of focus should be apparent as a cream or pinky-white glow. Getting the image onto the CCD chip is tricky because the chip is such a small object, at most a few millimeters in width and length, and so only capable of "catching" a field of view an arc-second or two across. Apart from the Moon and Jupiter, most astronomical targets are simply too small and faint to be particularly obvious when not dead center in the field of view and out of focus.

Once the image is on the CCD chip and focused, take a closer look at the computer screen. What will be immediately obvious is that even when sharply focused the image will seem to boil and blur. Two things can cause this. The first being thermals inside the optical tube, and allowing the telescope to cool down before using it will alleviate this. The second reason is atmospheric turbulence, or *seeing conditions*. Aiming the telescope at objects high in the sky is one obvious way to reduce this, by cutting down on the amount of atmosphere the light needs to pass through atmospheric turbulence will have a milder effect. However, unless you have great seeing with nice steady skies, chances are there will always be some atmospheric turbulence evident, and this is where webcams really come into their own. By recording a sequence of images you have the option to choose just the best ones and then stack or otherwise process them to enhance the detail still further; other devices, such as digital cameras, don't allow you to do this so easily.

Once you're happy with the focusing start playing with the telescope's controls. Move the telescope up and down (or in right ascension and declination), and see the effect on what the webcam sees. Most probably, you'll want to use the slowest possible settings; anything faster and the Moon will zip right out of the frame. Next, slot a Barlow lens in between the webcam and the telescope, and see how it performs at the increased magnification. Although you probably won't want to use a Barlow much with the Moon where the details are big and obvious, with the planets, especially Mars and Saturn with their small angular diameters, a Barlow will be essential just to get them big enough to see clearly on the frame. Try the Barlow out with a star diagonal, if you have one – a Barlow before a star diagonal offers about half as much magnification more as when placed in the usual position after the diagonal. If you have several Barlow lenses, try stacking them to increase magnification still further. This will obviously make centring and focusing more critical and turbulence more apparent, but it will make detail on the planets easier to record by casting the image over a larger portion of the CCD. Don't worry if the image is faint: up to a point stacking the images will compensate for this, allowing you to get brighter, more detailed images than you'd expect.

Now that the webcam is working, the next stage is to optimize the image capturing software. Exactly what applications you use for this will depend on what operating system and webcam you are using. As a rule, webcams come with Windows software at the very least, and some with Mac software as well. Links to web sites offering Mac and Linux software for unsupported webcams is in Appendix 1. An important step to using any webcam successfully is to switch off the automatic settings on the driver or image capture software. Usually these maximize the performance of the camera in bright light conditions and when the objects recorded occupy most of the frame. Neither of these conditions hold true for astrophotographical use. Another factor is that many webcams have infrared filters built into the lens assembly, and when this is removed, as is the case when a webcam is being used at prime focus on a telescope, the detectors that try to automatically set color balance don't work correctly. Instead, manually set things like exposure and color balance until you are happy with them. Finally, automatic gain can increase the noise to signal ratio in low light conditions, resulting in grainier, speckled images than normal. Remember that what you're after is not necessarily a bright image but a clean one: all else being equal, stacking can make up for overall dimness but it won't so effectively suppress noise. Which settings work well then? This will vary depending on your camera, the telescope, and what you are imaging. In general, it is best to switch off the automatic gain and exposure, but you will need to experiment with all the others to see what works best with your telescope and webcam combination. The Moon will work pretty well even with the default settings, but the planets and stars are more tricky, and with these resist the temptation to turn brightness up, and instead try to get the cleanest image you can with as little background noise as possible. Finally, make sure you switch off compression, if you can. With USB 1.1 webcams, some compression may be unavoidable if you choose too high a frame rate, but at low frame rates, you should be able to switch to "raw" or "uncompressed" image settings. With webcams using FireWire or USB 2, you should be able to use set the camera to send uncompressed images to the computer without any problems. Avoiding compression allows you the highest detail possible to be included in every frame, the downside is of course much larger file sizes: even short movies of a few tens of seconds can easily run to around 100 MB. But this inconvenience will be worth it later on when you are stacking images and trying to get as much detail into your final images as possible (we'll look at the details of taking pictures of the Moon on the one hand, and stars and planets on the other, a little later in this chapter).

The Principles of Successful Webcam Astrophotography

One you have your movies recorded, there are three basic steps to turning the sequence of short-exposure frames into a single image with more obvious color and detail: *registering*, *stacking* and *processing*; each step requires specialized software and techniques. Obtaining the software and learning how to use it is only half the battle though, actually being successful and producing good images takes a great deal of practice and naturally also depends on recording decent movies to work with in the first place. None of the software described here works miracles: it really is rubbish in, rubbish out, as computer programmers are wont to say! In this section, we'll go over the key steps involved in turning webcam movies into nice photographs, and the sorts of applications required to accomplish them. Fortunately, there are many options available, from high-end commercial packages through to modestly priced shareware, freeware and open source software from the Internet. Your choices will depend on the computer you're using: Windows has been the platform of choice for CCD users, and very largely for webcam astrophotographers as well, but there are some Macintosh and Linux alternatives. Appendix 1 includes links to suitable applications for all three platforms.

While wandering about the Internet, you will probably see some amazing pictures produced by many amateurs, seemingly not far short of Voyager or the Hubble Space Telescope. Your own initial attempts might seem to fall well short, but don't be discouraged! For one thing, steady skies if you're to record great movies to begin with as is the aperture and quality of the telescope used. Also, bear in mind that however good these factors are, you can only record images with as much resolution and detail as the aperture allows, just as with visual observing. In other words, a 90-mm Maksutov is never going to more detailed pictures than a 250-mm Newtonian. Practice and experience are *very* important as well; but that said, certain tips and techniques will help you get the good images from the very start, and that's where the rest of the remaining topics covered in this chapter come in.

Table 6.2. A summary of the main steps in successful webcam astrophotography

	Required hardware	Aim	Key techniques	Possible software
Recording	Webcam, laptop, telescope	Capture webcam images and store them on the computer's hard disk	Accurate tracking and alignment of telescope followed by careful aiming and focusing	No additional software required, since video capture software is usually supplied with webcam
Registering	Desktop or laptop computer	Align frames with one another so that the displacement of image effects of vibrations and atmospheric turbulence are cancelled out	Identifying the borders of targets (e.g., planets) and using these to manually or automatically nudge frames into alignment	Photoshop, Paint Shop Pro, The GIMP (small number of frames) or AstroStack, Registax, Keith's Image Stacker (for many frames)
Stacking	Desktop or laptop computer	Compounding image details from multiple frames into a single one with greater contrast	Selecting the sharpest frames and choosing enough of them to make the stacking process work well	As above
Processing	Desktop or laptop computer	Enhance the final image by tweaking contrast, colors, brightness and other parameters	Use of various filters such as unsharp mask, Laplacian filters and blurring; a key aspect is knowing when to stop applying filters	Primarily Photoshop, Paint Shop Pro, The GIMP; some tools also available in AstroStack, Registax, Keith's Image Stacker

Step 1: Registering the Frames

We tend to think of our eyes as cameras and the nerves and brain together as something like the film, capturing the image. In fact, our visual system is very different, and vastly more complicated: the eye sends *sensations* to the brain, and only there are they turned into the image we *perceive*. Take a look at a movie recorded by someone walking with a camcorder: the image moves about wildly; but when we walk about, our view of the world is much more stable. Likewise, when we look into the eyepiece of a telescope, at low powers at least, we hardly notice the ambient vibrations moving through the optical tube and the tripod. However cameras, including webcams, do not work this way, and as soon as you plug in a webcam this will become obvious: the image will seem to wobble. Motorized tracking will keep the target somewhere in the center of the frame, but not much can be done about atmospheric turbulence and accidental vibrations. Consequently, if you simply stacked all the frames in a movie without neutralising these small-scale shifts in the position of the target relative to the entire frame, the result would be blurrier image than any single frame (see Figure 6.5). Obviously, this is the reverse of what we're trying to do, namely to stack the frames in such a way they enhance the overall level of detail in the final image. To keep this from happening you need to *register* the frames so that the object of interest is in the same place on all the frames (see Figure 6.6). This means nudging frames a few pixels in one direction or another to compensate for the

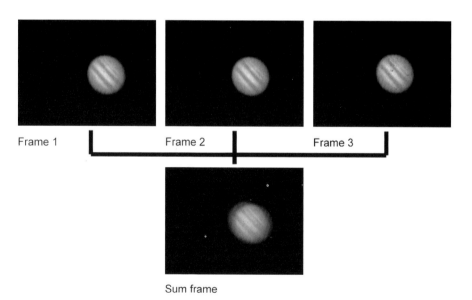

Frame 1   Frame 2   Frame 3

Sum frame

Figure 6.5. If three images are stacked simply by aligning the edges of the frame, as here with three shots of Jupiter, the final image is blurry because the position of target within each frame is likely to be different thanks to atmospheric turbulence and vibrations in the telescope.

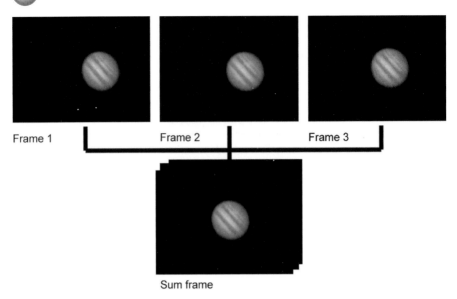

Frame 1   Frame 2   Frame 3

Sum frame

Figure 6.6. In contrast to Figure 6.6, this time the final image was created by aligning the three frames with reference to the target, Jupiter, rather than the edges of each frame. The resulting image is much sharper.

turbulence or vibrations that moved the target away from center of CCD during the recording process.

There are two ways to do this. One way is to use graphics programs like *Adobe Photoshop* and adopt a manual approach, aligning each frame one at a time. *Photoshop* is the graphics programs used most widely by professionals, but it is expensive, costing more than a small go-to telescope like an ETX 90! Fortunately, there are lower-cost alternatives, including the very popular *Paint Shop Pro* (Windows) that retails for around a hundred dollars, and two freeware graphics programs *The GIMP* and *JImage* (both Windows, Mac and Linux). *The GIMP* is an X Windows graphics application, and so runs natively in any UNIX-based operating such as Linux or Macintosh or in emulation on a Windows PC using something like Cygwin as mentioned earlier in Chapter 3. Luckily for those Windows users who don't want to run an X Windows environment on top of their operating system of choice there is a Windows-native version of *The GIMP*, but it still needs a bit work to configure and install. Despite the effort needed to get it up and running, *The GIMP* is very powerful, easily a match for *Photoshop* in many ways, and relatively straightforward to learn and use. *JImage* is a Java-based application that will run on any computer capable of running Java programs (pretty much anything running a modern operating system). It is widely used by scientists for image processing and analysis and although it isn't exactly a match for *Photoshop* slewed as it is towards a different user, it will allow you to do many of the key jobs like joining frames easily. Unlike *The GIMP,* it doesn't require an X Windows server or any complex adjustments to the operating system and in use feels much like a regular Windows (or whatever) program.

This is easy enough when you have a small number of frames but becomes steadily less fun when the number of frames increases much above a couple of dozen. We'll look at the exact technique in detail in the section of making mosaics of the Moon, but for now a quick summary will do. First, create a new graphics file with vertical and horizontal dimensions adequate to include the frames from the webcam movie (typically, 640 pixels across by 480 from top to bottom, though this may vary depending on your webcam). The format of the new document is important as some formats, like JPEG, don't allow you to save *layers*, and these are useful. What are layers? They are simply a way of allowing you to leave each pasted-in frame separate from the others, so that you can move and manipulate it without affecting the others. Until you *flatten* the image, each layer will contain a single pasted-in frame taken from the movie, the first layer the first frame, the second layer the second frame, and so on. Normally the default file format of the graphics application will accommodate layers (so with *Photoshop*, the .PSD file format is the one to use).

Now, copy and paste a nice frame from your movie and paste it into the graphics file. This will be *layer 1*. Copy another frame from the movie into the graphics file but as a new layer, *layer 2*, above the first one. Make sure that they are separate layers; what you don't want to do is paste all the frames into a single layer as this overwrites whatever you'd put in previously. Your graphics program will have a tool for adjusting the *transparency* of each layer; in the case of *Photoshop*, it is a slider called "opacity" on the Layers palette. If you increase the transparency of the overlying layer (i.e., decrease the opacity), then the one underneath will become visible. Now that you can see both frames at once, you can use the arrow keys to nudge the uppermost frame up and down, or side to side, until all the details in it align with those of the frame underneath. Then take the transparency of the upper layer back to normal, and repeat the process on a third good frame pasted into a new layer, *layer 3*, above the other two. Carry on doing this with all the frames you want to create a multi-layered image containing a number of perfectly aligned frames. This multi-layered image can then be stacked, or *averaged*, and then processed in the ways described below.

A second way to register frames is to employ software designed especially for the job, as with *Registax* (a Windows program for aligning BMP and AVI frames) or *Keith's Image Stacker* (a Mac program for QuickTime movie frames). These automatic registration programs work in the same basic way. First, the user singles out part of an image that the computer can use for alignment, for example the disc of a planet, on one frame. Then the computer uses that information to align all the other frames with the first (the details are covered in the section on imagine stars and planets later on in this chapter). Manual and automatic registration both have their advantages in certain situations. Obviously automatic registration is the preferred choice if you've recorded a movie that includes hundreds of frames, but on the downside, it is very sensitive to the quality of the individual frames. If there are very many blurry frames, these can mess up the alignment process dramatically, so it is vital to remove the bad frames as much as possible to begin with, and only align the good frames. There is also a bit of trial and error involved in getting the hang of selecting the portion of the frame that will provide the basis of the alignment process; the smaller and sharper the detail, the better the alignment and the less the degree of error likely to occur. On the other hand, manual alignment works especially well for images of the Sun and Moon, where typically you are working with a few frames, perhaps less than a dozen. Here, registering the frames by eye is easy and probably more effective as you want to align lots of different craters and other structures all across the frame, and not just one small part of the frame. Automatic alignment isn't much use for creating mosaics, either, where frames only overlap at the edges and not completely.

Step 2: Stacking the Frames

Once the images are registered, the next step is to stack them. This is the step that really seems magical, if done properly it enhances brightness and contrast, and detail can become much more obvious. But the first few times the process can easily become disappointing: stacking doesn't bring out detail that wasn't there to begin with, and as we've seen, if the images are not tightly registered then stacking messes up the details rather than enhances it. As with registering the frames, both manual and automatic approaches exist. In the case of a mosaic of the Moon or Sun, then stacking manually can mean nothing more than adjusting the colors of each frame so the joins between them are hidden, and then flattening the image to make the single mosaic image. Manually stacked frames of planets or stars will need to be averaged rather than flattened since you want each frame to contribute detail to the final image rather than overlap it, as is the case with a mosaic. Again, graphics applications like *Photoshop* will allow you to manually align and stack images easily and the details on doing this will be gone into in the section on imaging the Moon that follows shortly. For automatic stacking, there are numerous programs available for Windows and Mac computers, such as *AstroStack*, *Registax* and *Keith's Image Stacker*. (As a bonus, some of these programs also include tools for processing images too.) Typically, these programs allow the user to import a webcam movie, such as an AVI or QuickTime movie, and then give the user the option of including or discarding particular frames from the final stacked image. The included frames are then registered and then stacked, and then the resulting composite image saved to disc for subsequent processing.

If you have a lot of noise in each frame, you can also *subtract a dark frame*. This isn't precisely what its name suggests, a black frame, but one taken by the webcam with a piece of opaque card or something placed in front of the lens so that it isn't actually recording anything. What this does is to allow you to record the background noise inherent to the webcam, in other words the speckles of light that are scattered across all the frames whether or not the webcam is imaging something or not. You can then use some of the programs mentioned here to take away this background noise from all the frames containing whatever you're imaging. This can result in a cleaner, noise-free set of frames to work with. Subtracting dark frames is most useful when working with long-exposure images though, as would be possible with a modified webcam used for deep sky imaging, rather than short-exposure imaging of the Moon and planets with an unmodified webcam.

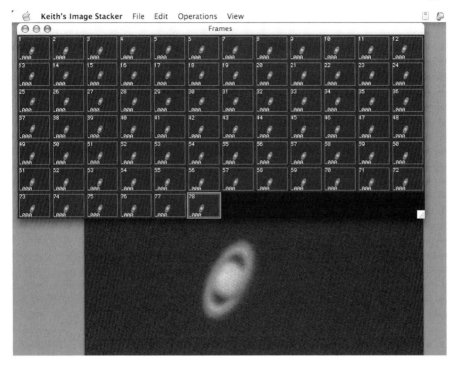

Figure 6.7. All image-stacking programs work in the same basic way, allowing the user to create a "stacked" composite picture based on the original webcam movie but excluding any fuzzy frames that would obscure rather than enhance detail. This is *Keith's Image Stacker*.

Step 3: Processing the Image

This part of the process is the most fun, which is lucky as there really isn't an ideal recipe, only basic ideas for experimentation. The only thing that can go wrong is over-processing, so it's always a good idea to work on a copy of your stacked image rather than the original. There are two processes involved: *sharpening* and *blurring*. Despite its name, the best sharpening tool is the *unsharp mask* tool included in many graphics programs for processing photographic scans (Figures 6.8). It happens to work exceptionally well with astronomical photographs too, and so you will find an unsharp mask filter of some sort built into the stacking and registering programs mentioned earlier. What unsharp masking does is to increase the contrast between light and dark pixels wherever they meet. Because our eyes perceive edges best where there is a strong difference in contrast, this makes the edges of a graphic processed using an unsharp mask tool seem more clearly defined. Astronomers often talk about "contrasty" images and this is what they mean; essentially, the dark regions of the object being observed seem darker compared with the lighter regions, which seem lighter. The unsharp

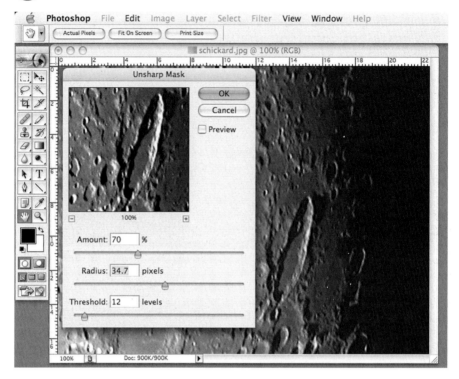

Figure 6.8. This is an example of the Unsharp Mask filter at work on a photograph of the Moon centered on the crater Schiller. The original image is in the background while the floating Unsharp Mask window shows how the image will look once the filter is applied. Among other things, note the sharper boundaries between the shadows of the crater's rim and its floor, and the contrast of the light catching the rim on the opposite side with the shadows beyond.

mask is an artificial way to increase this contrast. The danger is in doing too much the image ceases to look natural and looks obviously artificial instead (Figure 6.9).

There are three settings involved with the unsharp mask: amount, radius and threshold. The *amount* is the intensity of the unsharp masking. In *Photoshop*, for example, a slider lets you choose from 1% to 500%, i.e., from one-hundredth of the normal intensity of unsharp masking through to five times the normal amount. To begin with, 100% is probably fine. The next setting is *radius*, which is a measurement of the "sharpening halo", i.e., the size of the area that will be sharpened. So a one pixel radius will have the filter sharpening up to one pixel outwards from the light-dark boundary being worked on (these are chosen by threshold, which we'll come to in a moment). Usually, only a little is needed, and although the *Photoshop* slider offers values from 0.1 pixels up to 250 pixels, it's best to start with values of 1 pixel or less. Incidentally, it is usually better to set

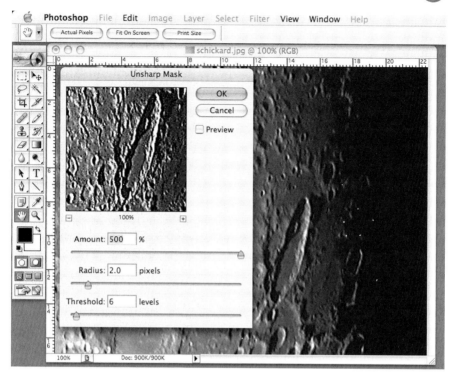

Figure 6.9. Excessive use of the Unsharp Mask filter will result in unrealistic images. Whereas the preceding picture showed an example where the filter improved the original, here the resulting image has lost its smoothness and the contrasts between light and dark are far too extreme; the picture looks more like a bad photocopy than good photograph.

the radius *low* and *increase* the amount as required, rather than the other way around. Too much radius simply blurs detail away, and often throws up haloes around the edges of the object as well. Finally, *threshold* is the setting that determines what degree of contrast between light and dark is worthy of being unsharp-masked. The lower the value, the less stringently the filter excludes contrast boundaries, and the more of the image is processed. A higher value only allows the strongest contrast differences to be accepted (i.e., very dark against very light regions), and so less of the image is processed. Obviously, if you set the threshold too high then nothing gets through and no sharpening will take place; but set the value too low, and background contrast boundaries will be enhanced as well as the detail you want, and the image will become noisy and artificial looking. A threshold values from 0 to 10 work well, though you may need to use higher ones in some cases.

A second sharpening tool that has become popular with astrophotographers is the *Laplacian filter*. This mathematical trick compares each pixel with the one above, below, to the left, and to the right, and then processes them accordingly. In

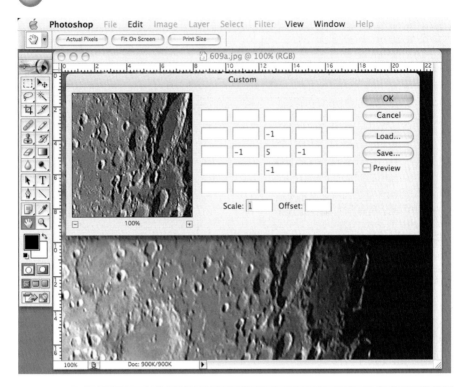

Figure 6.10. Many advanced astronomers find that Laplacian, or convolution, filters are preferable to Unsharp Mask filters for improving images.

a *3 × 3 convolution*, as this comparison is called, only the first pixel in each direction will be used (fitting inside a box three across and three up, hence the "3 × 3"). If two pixels in each direction are used, the convolution is 5 × 5, if three pixels, then 7 × 7, and so on. Specifically, the filter multiplies the brightness by some preset value, typically increasing the brightness of the selected pixel and decreasing the brightness of those immediately adjacent to it. Some astrophotography processing programs have the Laplacian filter built into them, while graphics applications like *Photoshop* allow you to create one easily enough using the *Custom Filter* (under the Filter menu item in the Other section). The Laplacian filter can work wonders, but as with the unsharp mask it is easy to ruin an image by using it too heavy-handedly. If you do create your own Laplacian filters within *Photoshop*, what numbers work well? Begin by multiplying the selected pixel by a small positive value (e.g., 5) and the ones adjacent to it by a smaller negative one (e.g., −1) and see how your image comes out (Figure 6.10). Save a copy of the filtered image, and then nudge these values up and down until you find the right combination of values for your work. It may well be that the values that work best on images of the Moon don't work so well on Jupiter, and vice versa. Laplacian filters can over-sharpen images by a considerable degree, in which case some *blurring* will be necessary.

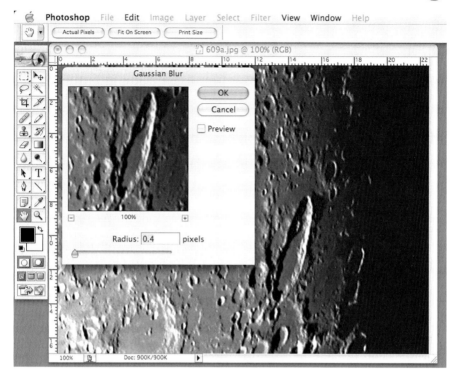

Figure 6.11. Blurring an image is normally the final manipulation, since it "throws away" data from the image. When used very carefully, a tiny amount of blurring can make the difference between a nice, sharp image looking vaguely artificial or impressively realistic. In this example, a Gaussian Blur filter has softened the edges of the contrast features, making the image more pleasing.

Blurring sounds like a strange thing to do after going to all this effort to sharpen the image, but used judiciously it can be a very useful step. Why use blurring at all? It is a way of removing noise but leaving the detail behind. If you've used unsharp mask or a Laplacian filter to improve the detail (such as the belts on Jupiter), then a bit of blurring could be just the ticket to get rid of any artificial noise that has cropped up elsewhere in the image, such as across the dark sky around the planetary disc. The *Gaussian blur* tool is particularly popular among astrophotographers, and is sort of an opposite to the unsharp mask, meaning that any contrast below a certain threshold will be blurred away (Figure 6.11). In *Photoshop*, there is only a single variable parameter, the *radius*, which is the maximum number of pixels processed at once. The bigger the radius, the more blurring you get. Because the Gaussian blur throws away detail, this is a filter to apply only at the very end, once you are happy that the image otherwise perfect. If you blur the image too much, the details are lost completely, so be sure to work on a copy (Figure 6.12).

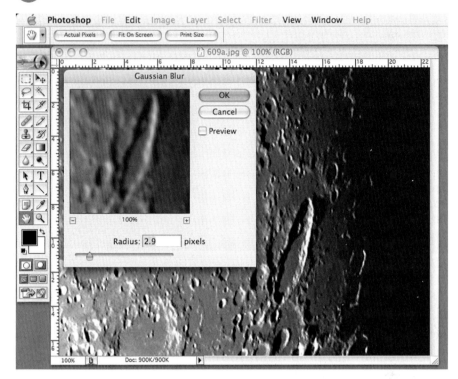

Figure 6.12. Too much blurring obscures detail, as shown here. For this reason, it is essential to work on a copy of your image: once image detail has been blurred away, it can't be brought back!

Putting the Principles into Practice: Imaging the Moon

It's time to get down to specific projects. The Moon is an ideal object to being with because it is big and bright, and even if not in focus it should be simple enough to get the image onto the CCD chip. If your telescope tracks the Moon as well, then the job of keeping the image centered is easy, but with a bit of care it is even possible to track the Moon manually using slow-motion controls on manually operated mount. I've used an alt-azimuthally mounted telescope to take pictures of the Moon by nudging the telescope up and across as required, all the while recording the view with my webcam. The other great thing about the Moon is that instead of forcing you to work with small images that have you eking out scrap of detail from a planetary disc a hundred or so pixels in each direction, with the Moon your canvas is much, much larger. Indeed, as any one frame will only cover a small piece of the Moon's surface, it's probable that after stitching all the frames together you'll want to shrink the final image to make it much more manageable. With luck, resizing the image will make any little errors much less noticeable, and even if you only use a single frame for any given bit of the Moon, there should still be plenty of detail to make for an impressive and satisfying image. In this section, we'll work stepwise through the process and look at how to make a Moon mosaic, the best way to create an image of the whole Moon using the relatively narrow field of view inherent to the webcam and telescope combination.

Although you can image the Moon in twilight, the lack of contrast makes this situation a bit less desirable than imaging at night. Even so, I'd recommend starting with a relatively young Moon though for one very good reason: there is less of the Moon to image, making the process quicker and easier because that if with a half or gibbous Moon, since you will need to record fewer movies to cover the visible surface. A four- or five- day- old Moon is ideal. However, you also want the Moon high enough above the horizon for the seeing to be good. At certain times of the year, such a young Moon will tend to be closer to the horizon than at

Figure 6.13. Although imaging the Moon is the easiest project to begin with, it is still dependent on steady atmospheric conditions and a steady telescope. Even if these criteria are met, seeing conditions will vary from moment to moment, making it essential to capture sufficient webcam footage to be able to cull blurry frames (such as that on the left) while retaining the sharp frames for subsequent manipulation and use (like the one on the right).

others, and it may be that an early morning session with an old Moon would make better sense. For example, in the northern hemisphere, during the autumn the waxing crescent Moon in low in the sky but high in the sky during its waning crescent phase.

Producing Sharp Images of Lunar Features

Although many amateurs are most interested in creating images of the entire Moon, some of features like the Straight Wall, and of course, craters like Copernicus and Clavius, are dramatic enough to warrant portrait shots of their own. In this case, extract a few good frames and stack them exactly as you would shots of the planets, as described in the section on that topic. You don't need very many frames for this to work well, we're talking ten to twenty rather than the hundreds used to get the best possible images of Jupiter or Saturn. Aligning the frames will be important though, particularly if you used an unmotorized mount and let the Moon drift across the field of view. Luckily, with so much detail visible aligning the frames is easy either manually or automatically. Since we'll be looking at automatic stacking in the section on planetary imaging below, we'll concentrate here on manual stacking using a graphics program like *Photoshop* which works very well with situations where you are stacking just a handful of images.

The first thing to do is to register the frames using the manual registration process described previously. Let's say you have identified five nice and sharp frames from the movie you recorded. Copy the first two frames in as separate layers. Reduce the transparency of the one on top, and nudge it a pixel at a time until it aligns perfectly with the one below. Return the transparency of the top layer to 100%, and then paste in the third frame and repeat the process. Continue this until you have aligned all five frames. Next comes the actual stacking part, but before you do this, it is worth making a copy of your multi-layered image file and working on that, just in case something goes wrong. This is great advice with any graphics project – save early, and save often! Open the *copy* of your stacked image and make sure all the layers visible but change their opacity so that it is equal to a hundred divided by their position in the stack: i.e., the first layer has an opacity of 100/1 = 100%; the second 100/2 = 50%; the third 100/3 = 33% and so on. After you've done this, either *flatten* the image (in *Photoshop* this is under the Layer menu item) or *add* the frames together (using the Add option within the Apply tool found under the Image menu item). Now is a good time to save the results. You will probably need to crop the image a little since the frame alignment process is bound to leave some of the edges containing information from fewer frames than the central part, and a little sharpening may improve the results still further. While stacking frames can help overcome the vagaries of seeing and atmospheric turbulence, even a single good frame of lunar features can be surprisingly impressive. As with viewing the Moon at the eyepiece, the illumination of the features makes a huge difference, things close to the terminator having much better contrast and definition. Moon-mapping software can be very useful in this regard for helping you identify the best nights to catch the best views of specific objects.

Creating Mosaic Images of the Whole Moon

Using webcam movies to make a composite image of the Moon is surprisingly easy; the real trick to imaging the Moon is making sure you capture the entire surface in one session! This is simpler with a motorized, where using the up and across motors it should be possible to scan the entire surface from top to bottom, but with an unmotorized telescope things are a bit trickier. The best thing to do is to start from the top, work you way down a bit at a time, and err on the side of caution. It is much better to go over the same bits of the surface twice than it is to leave a bit off and find that your mosaic has a black patch in the middle that you forgot to photograph! A second issue to bear in mind is that as the Moon travels across the sky it rotates, and so if you take too long perceptible changes in the orientation of the later movies compared to the earlier ones become apparent. If you get everything done in half an hour then this shouldn't be too much of a problem, but much more than this and you'll soon notice that frames don't align perfectly. Equatorially mounted telescopes won't have this problem, of course, because the telescope will rotate at the same rate as the Moon, but with alt-azimuthal ones, be they manual or computer controlled, it is well to prevent this problem become serious by working quickly.

You may be able to record a single movie that covers the entire Moon in a single go with a motorized telescope that will allow you to sweep down and across the surface easily using the default frame rate. Manual telescopes are trickier, and a better approach with them is to reduce the frame rate (to five or ten frames per second), lock the telescope and let the Moon drift by instead. When one longitudinal "fly-by" is complete, bring the Moon back into view, lower the telescope a bit so it now sweeps a lower section of the surface, and again, let the Moon drift by. In this way, you can get a much smoother movie than is possible if you manually try to move the telescope down and across. By keeping the frame rate low enough these movies won't be unmanageably huge even if they last several minutes, as they would be at the default frame rates of around thirty frames per second. Incidentally, you don't need to record the dark side of the Moon except perhaps when it is very young (or old) and the earthshine is enough to illuminate the dark side of the Moon sufficiently well for the webcam to image it. Outside of these times the dark side will be black, the contrast between the dark and light sides of the Moon so great the webcam can't image both at once (any more than our own eyes can); so at these times simply record that part of the disc that is illuminated.

Having assembled a bunch of movies resembling a film taken from the Apollo command module orbiting the Moon, you're now ready to begin taking individual frames from the movies and stitch them together to make a single big mosaic. To do this you'll need to use your graphics programs, be it *Photoshop* or whatever. Begin by creating a single large empty document and fill it so that it is white not black (in *Photoshop* you should have the option to fill the image when you create the new document, otherwise choose Fill from the Edit menu). It needs to be white so you can see the gaps between frames clearly, so although black would seem more logical, it actually isn't a good choice. How large the document needs to be depends on how much of the Moon fits into a single frame, but I find starting with something ten times the size of a single frame is usually enough to begin with. You can always make the document bigger or smaller as required later on. It doesn't really matter where you begin, but if you are doing an image of a crescent Moon, then copying a frame that includes the top or bottom points of the crescent makes sense, giving you a nice obvious place to work from. I find working from an edge inwards (like you'd do with a jigsaw puzzle) by far the easiest approach, but it really doesn't matter. What will be obvious after you've pasted in the first frame is that this document will be *big*. Exactly how big will depend on the field of view covered by a single frame, and this is a function of the focal length of the telescope, the magnification of the Barlow lens (if used), and the size

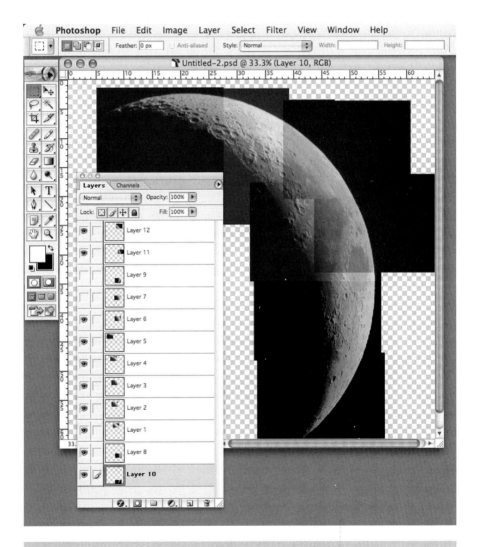

Figure 6.14. Each frame is pasted into the mosaic as a new layer, and that way it is easy to adjust the color and contrast of each frame so that it matches its neighbors perfectly.

of the CCD chip inside the webcam. For example, it only takes about five whole frames to cover the width of the full Moon using my iBot FireWire webcam when used with a Barlow lens and a 76-mm f/6.3 refractor. However, it takes twice as many frames to cover the same width with my Logitech USB webcam and 200-mm f/10 SCT without a Barlow lens.

As you add more and more frames, with luck you should start seeing the picture looking increasingly like the Moon and not just small pieces of it. Each new frame is a new layer, so more than likely the final image will include dozens of layers. When you paste in the frame, you'll need to move it around to get it lined up with the first frame. Normally, you can either move the frame by "grabbing" it with the cursor or use the mouse to move it about, or else nudge the thing up, down or across using the cursor keys on your keyboard. I find it best to use the mouse to move the thing roughly into position and then magnify the image by a factor of two or three times and use the cursor keys for a final adjustment. You will quickly find that it is easiest to align images that have a significant amount of overlap between them. Find a crater or some other small but sharp feature they have in common and use that to align the second frame with the first one. In most graphics programs, you can toggle the visibility of a layer on or off (in *Photoshop* for example you click the eye icon beside the layer's name and preview icon in the Layers palette to do this). By doing this you can blink between the first layer and the second, and see how far you need to move the second layer to align the frame it contains squarely with the first. With each new frame pasted in as a new layer, you repeat this process until the composite image is complete and seamless.

Even though you might have a nice big picture of the Moon, an obvious flaw will become apparent at once: the colors and brightness on each layer will be different. The document will look more like patchwork quilt than a single, big image. This is especially true as you get towards the edge of the Moon, where part of the frame includes black sky. Because of the way the hardware in the webcam works, where part of the frame is dark and part bright, it changes what it sends to the computer to try to even up the overall brightness of the image, in its way trying to be helpful. While that is useful enough when using a webcam for video-conferencing, it isn't what you want in this situation, and that is why you have to start manipulating each layer to get rid of differences in brightness between frames. The first thing to do is discard the colors. There isn't much reason to keep colors in a picture of the Moon since shades of gray dominate almost entirely, and discarding color gets rid of some of the difficulty in getting each frame to match. You can do this by *desaturating* the image (in *Photoshop* this is an option in the Adjustments section under the Image menu). A grayscale image also contains less data and is therefore smaller, so it takes up less space on a hard disk. Finally, the attractiveness of the features on the Moon comes largely from the contrasts of light and shadow, and these are things our eyes see particularly well in black and white images.

With the color discarded, you'll find a variety of tools to work at the next stage of processing. These include the *brightness* and *contrast* sliders, and tools for changing less obvious things like *curves* and *levels*. Brightness and contrast are obvious enough properties, and playing with the sliders will show you what they do. Curves and levels are related, and are ways of adjusting the range of tones in an image. If you open up the curves tool in your graphics program, some sort of graph with a straight line running through it will appear. This line shows the relationship between the input tones on one axis (the horizontal axis in *Photoshop*), i.e., those recorded by the webcam and saved with the file, and the output range on the other (the vertical one in *Photoshop*), i.e., those shown on the screen. Pull the curve towards the top left and the picture will get brighter as brighter tones replace the original tones, and pull the curve to the left and the opposite happens. Unlike simple brightness and contrast changes the curve can be adjusted in a more complex way, with the curve being bent into an s-shape for example, making the dark parts darker and the bright parts brighter but leaving the middle tones alone. The levels tool does a similar thing to this s-shaped curve, and can be a very useful way of enhancing the overall contrast of an image. With a grayscale image the way these tools works will become second nature quite quickly (they are much less simple when used with colored images). Nonetheless, it is always a good idea to save often and not to work on originals but on copies. That way, if you make a mistake, you can replace the messed up file with another copy of the original.

Figure 6.15. Adjusting the brightness and contrast is one of the essential steps to processing webcam images. This shot of the crater Clavius shows the effect of the Curves tool in *Photoshop*. On the left is the original image, in the center with the image mapped to brighter shades by pulling the curve towards the top left, and to the right is the result when the curve is pulled to the bottom right.

There are two actions required to complete the mosaic. The first is to nudge the frames to improve the alignment, by finding features that span two frames and then ensuring both halves line up properly. The second thing is adjusting the colors and tones so that the disjunction between adjacent frames isn't apparent. There is a huge amount of trial and error in getting this second action done; but begin with brightness and contrast, and then move on to curves and levels. Once the lunar disc is complete, the next things to look at are the spaces all around it. These will be a mixture of black sky and chunks of the white background created with the document when you began. Fill these in one of two ways. One method is to simply create a new layer behind the layers containing the shots of the Moon (in *Photoshop*, choose New under the Layers menu), and then fill this with black or dark grey (again in *Photoshop*, use Fill under the Edit menu item). With luck, this will hide the empty space nicely, but if the sky or non-illuminated parts of the Moon in the frames don't fade to black or the shade of dark grey you used for this background fill, then a join could be obvious. The second method is a bit more complex but allows a much more controlled fit between the artificial sky background and the dark parts of the frames taken with the webcam. Because you need to flatten the image (in *Photoshop*, using the Flatten Image command under the Layers menu item), you want to be working on a copy of the image. Once flattened you loose the layers, which means if you need to go back and tweak things frame by frame, you can't. So, flatten a copy of the image to turn it into a single layer image. Now choose the *magic wand* tool (this will be on the floating palette in *Photoshop* and most other graphics applications, though sometimes there will be a keyboard shortcut to activate it too). You'll notice that one of the options available is *tolerance*, normally given as a percentage. The magic wand chooses adjacent pixels that match, within limits, the one clicked on, and the higher the tolerance the bigger the difference in color and tone chosen. Therefore, a low tolerance will only choose a region matching exactly the selected pixel, while a high tolerance will allow pixels that differ greatly in tone and color to be included. Try it out on the white background and you should see it selects all the empty spaces (if your composite image divides the background into two or more non-adjacent regions, you can select more than one region with the magic want tool, in *Photoshop* by shift-clicking). Now with the background selected use the brightness and contrast sliders to darken the region to dark grey or black as you prefer (take brightness down and contrast up). Adjust them to get a close match to the dark regions on the frames. Finally, use the magic wand tool a second time clicking on the dark sky but keeping an eye on the tolerance level. Begin with a moderately high tolerance, say 30, and you'll see that this time not only is the fake dark background chosen but so too is the dark sky in the frames and the non-illuminated part of the Moon. If the tolerance is too high, significant regions at the terminator will be included, what you want is the entire sky to be chose plus enough of the non-illuminated side of the Moon to blend with the terminator and the rest of the lunar disc imperceptibly. I find values between 10 to 30 work well, but a lot will depend on the contrast and brightness of your image and how sharp or otherwise the terminator and the limb of the Moon is. Once selected, you can again adjust the brightness and the contrast to make the selected region much darker. You can also use filters, such as blurring ones, if there are artefacts in the image you'd like to remove. If the sky recorded by the webcam still doesn't match

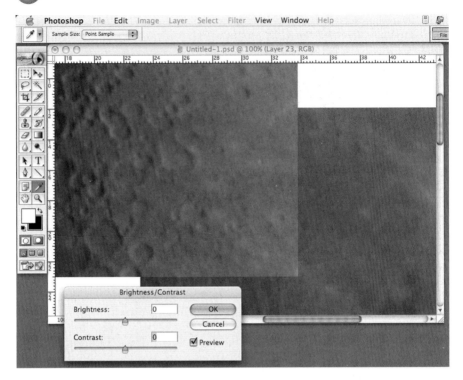

Figure 6.16. By toggling the visibility of adjacent frames on and off, it is easy to nudge individual pictures of parts of the surface of the Moon into alignment so that there are no gaps or discontinuities between them. In this way, a complete mosaic can be made, but the different brightness values of the frames recorded by the webcam remain obvious.

the artificial night sky fill, you can always clear the selected region and then fill it with black or grey.

One way or another you should now have a nice, contrast composite image of the Moon against a dark gray or black background. The final adjustments are ones to improve the presentation rather than the quality of the image. Usually rotating the image, so that the Moon is oriented with north upwards improves the image and makes it easier to identify craters and seas with features on a map; normally, the terminator will arc across the north–south axis. Very often, the Moon will be oriented with the axis of rotation leaning towards the left or right, especially if you took pictures of a very young or old Moon because these will close to the horizon and either setting or rising. In *Photoshop*, use the Rotate Canvas option under the Image menu item. Some telescopes will flip the image to horizontally as well, in which case using the Flip Horizontal command in your graphics program will bring it around to the correct orientation. Rotating the canvas creates new regions of empty space around the now-tilted original picture; fill these using the same techniques as before. You may also want to resize the image, or at least save a smaller copy of the image. Smaller images look sharper

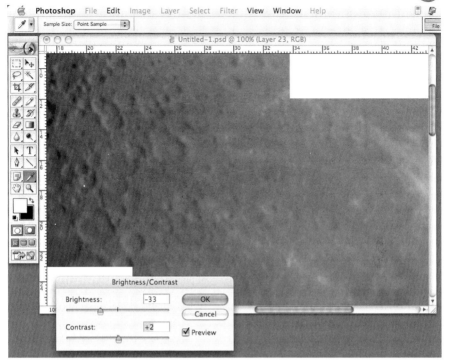

Figure 6.17. Careful use of the Brightness and Contrast sliders provide the easiest way to get rid of brightness differences between adjacent frames, at least for black and white ("grayscale") images.

for one thing because the blurriness caused by the atmosphere is most obvious with the fine scale features; lose these, and the whole things looks better. However, a smaller image is also better for display on a web page: try to get the whole image resized so it fits comfortably on the computer screen or in a web browser window. It the image is too big, and you need to scroll about to see the entire thing, a lot of the visual impact of a whole-disc image is lost.

One of the most effective improvements you can make is to add labels to your picture of the Moon. Some people like to put nothing more than the date and phase of the Moon onto the image, but others are more ambitious and label up the various features. Adding labels is done using the text tools in the graphics program, but one important thing to remember is that once the image is saved with the labels in place as a flattened image (as it has to be if you want the popular JPEG image format), you can't edit or remove this text. Therefore, it is best to add labels to a copy of the image. Put each labels into separate layers as well; that way you can change labels you get wrong, or move them about if you find that some of your labels need to be differently positioned for greater clarity. Although you can put the names on top of large features like the lunar seas, with smaller features like craters simply dropping the label on top can obscure them;

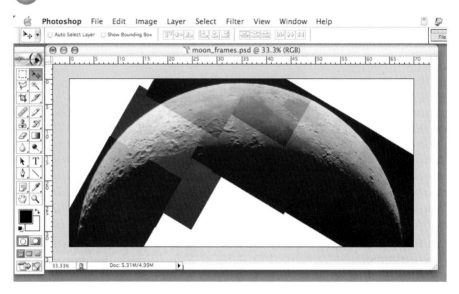

Figure 6.18. Here is a complete mosaic of the Moon before adjusting the brightness of the various frames to hide the joins.

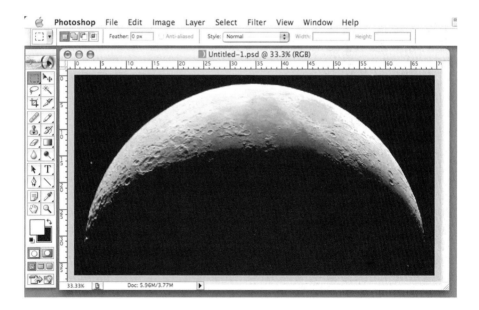

Figure 6.19. The same mosaic of the 4-day old Moon shown in Figure 6.18, but with the brightness of the frames adjusted and the background filled with black to recreate the night sky.

instead, mark the feature out somehow and put the appropriate label somewhere nearby. You could use arrows or lines, but I happen to like using the bullet symbol ("•"). It is small but obvious, automatically centered with the midline of the text, and unlike lines or arrows can be typed straight into the text field with the name of the object. Positioning the bullet before or after the name, and the judiciously use the space bar to separate the name from the bullet. This gives you plenty of scope to tweak the label so that you can get the bullet *on* the feature but the name *far enough away* it doesn't obscure anything. On a Windows keyboard the shortcut for the bullet symbol is *Alt+0149*, and on a Mac keyboard *Option+8*. Once done, save one copy of the image with the labels and a second copy of the

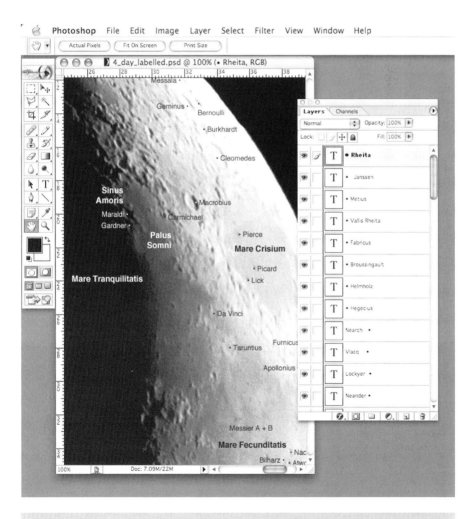

Figure 6.20. Adding labels is one of the best ways to enhance your pictures of the Moon. Add each label as a separate layer so that you can move individual labels if you have to without disturbing the position of the others.

image flattened $ ANOT °x lin JPEG format to share with others or put on your web site.

Imaging the Sun

In general , image the Sun in the same way as the Moon. Instead of craters, you can focus on sunspots by stacking a few good frames, or else you can create a composite image by pasting together frames from a series of movies that scanned over the entire surface of the solar disc. The big difference is that with the Sun you are working with a source of light so powerful it can seriously damage both your eyes and any photographic equipment you attach to the telescope. To make sure this doesn't happen you need to use a solar filter that covers the aperture of the telescope. There are two sorts: ones made of glass that fit onto the front of the telescope rather like the lens cap; and others that are made of a special plastic film, aluminum-coated Mylar, that attached to the front of the telescope. Either sort works well, the solid glass ones being more convenient but expensive, while

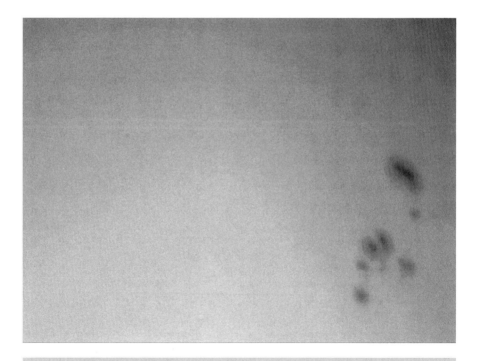

Figure 6.21. Although the Sun is an easy target for budding astrophotographers, as shown here with a webcam shot of some sunspots taken through a small refractor, the Sun is a potential hazardous object for both observer *and* webcam. Make sure you use a proper solar filter at the objective end of your telescope.

Mylar is very inexpensive but does require some handiwork to contrive a housing that fits the film onto your particular telescope. Mylar has a reputation for producing images with most detail and highest contrast, but the blue-grey cast to the image is unattractive. The remedy is to use color filters, such as Wratten #15 Deep Yellow or #21 Light Orange. You can usually screw color eyepiece filters into the webcam adapter, but if you are using a homemade webcam adapter take a piece of photography gel over the front of the adapter and tape it into place. Glass filters normally give the image a yellow or orange cast anyway, so the use of color filters is unnecessary.

The other catch to taking pictures by day is not with the telescope or the webcam, but with the computer. Laptop displays work notoriously badly in sunlight, particularly color ones (for some reason active matrix greyscale screens works rather better). A simple solution is to put the computer somewhere in the shade, and one way to do this is to use a large cardboard box, big enough to take the computer with the laptop screen opened up. You can easily make some small holes in the sides of the box for things such as the USB and power supply cables.

Once you have the images recorded, the rest of the operating is very much as with the Moon. Although there aren't any permanent features to label, it is very worthwhile annotating sunspots if they are present, in the same sort of way as with pictures of the Moon. The web site of the Solar and Heliospheric Observatory, or SOHO, (http://sohowww.estec.esa.nl) usually has a map of the current sunspots and their NOAA designations, updated daily, as do several other large terrestrial observatories around the world.

Moving Up to the Planets and Bright Stars

Compared to photographing the Moon, the planets offer the astrophotographers much more of a challenge. For one thing, they are small and relatively dim, making it difficult to center the image of the planet on the CCD and in focus. Secondly, you need high magnifications to record detail, which means that the telescope must be mechanically and thermally stable, the optics well collimated, and the skies steady. Finally, the small angular size of the planets demands high magnifications if you want to see details, and therefore automatic tracking is essential or the planet will simply zip out of view before you've had a chance to get a movie recorded. However, the principles are simple enough, and some hobbyists have gone on to produce some very striking images.

Of all the planets, Jupiter is the easiest to work with. It is large and bright, and the various equatorial bands provide a wealth of detail to try out your favorite photo enhancement techniques. The four Galilean moons are bright enough to show up on webcam movies taken with all but the smallest telescopes, as are their shadows during those times when one of them crosses the face of Jupiter. Saturn is almost as nice to work with, though much smaller and fainter, its redeeming feature being the spectacular ring system. Compared with Jupiter, the details on Saturn need higher magnifications, making this a planet for nights with steadier than average seeing. When that happens, your webcam can easily capture equatorial and polar banding, the Cassini Division, the elusive Crêpe ring, and perhaps even the moon Titan. In a good year, Mars is equally rewarding, and although small, it is bright and there is plenty of detail. Even with moderate magnifications, your movies will record the polar ice caps and Mars' distinctive albedo features like Syrtis Major. Any webcam will record the phases of Venus easily, and likewise with Mercury, though more magnification and steady skies are essential. Uranus and Neptune are both possible but dependent on large apertures and high magnifications, and in either case, all that is visible is a tiny colored disc with no detail of any sort apparent.

Recording the Best Possible Movies

The most important thing with imaging the planets isn't so much the technique used when processing the movies, but recording adequately good movies in the first place. Because high magnifications are used, atmospheric turbulence is exaggerated; consequently, if the planet is too low in the sky, you will end up with hopelessly blurry movies. At best you can hope to snatch a few sharp frames from the mass of bad ones, and stack those, but to be honest you are much better off aiming the telescope at a planet 45° or more above the horizon, where the sky is steadiest. You also want to optimize your telescope for high-magnification viewing, at the very least making sure collimation is spot-on and that the telescope is tracking smoothly and accurately.

A good quality Barlow lens makes a big difference if you are pushing magnifications towards the limit of your telescopes performance. Look for a full sized Barlow (as opposed to a "short" one) made with apochromatic lenses. Short Barlow lenses tend not to be so sharp, especially at high magnifications, and without apochromatic lenses false color will start to creep into your pictures. Meade, Tele Vue and various other companies make apochromatic Barlow lenses in a variety of sizes. Barlow lenses that double the magnification ("×2 Barlow lenses") are the most popular and are useful for both imaging and visual observing, and good one is an sound investment. Barlow lenses offering up to five-fold magnification can be useful, especially with short focal length Newtonians and apochromatic refractors that don't provide enough magnification with the ×2 Barlow. If you have very good skies you can use these Barlow lenses with traditional long focal length telescopes to make the image of Saturn or Jupiter as big as possible, but to begin with, you will probably find working with smaller images easier and more rewarding in the short term. Variable power Barlow lenses exist, but these tend to be inferior to single power ones (for much the same reasons as zoom eyepieces are at best a compromise). Tele Vue also make something called a "Powermate" that they market as an alternative to the traditional Barlow lens. They come in a variety of powers, from ×2 to ×5 in both 1.25 and 2-inch fittings. They are expensive: the ×2 Powermate costs almost three times as much as a good apochromatic ×2 Barlow; but the images they produce are very good indeed.

While on the topic of optimizing your telescope's performance, a much overlooked but important piece of the set-up is the star diagonal. Unfortunately, the star diagonals that come with even the higher end model refractors and SCTs

Figure 6.22. A good quality Barlow lens is an essential component to an astrophotography set-up. Expect to pay around $100 for one that will give you the magnification and sharpness you need, such as this Ultima Barlow from Celestron. Besides being great for photography, a decent Barlow is very useful for visual observing, too (photo courtesy of Celestron).

from Meade and Celestron are of middling performance. For a start, they are often of the prism rather than mirror design, and long Barlow lenses won't fit into prism star diagonals (when pushed in at the eyepiece end of the diagonal, the end of the Barlow will hit the glass prism). Moreover, generic prism star diagonals tend to scatter more of the light that strikes them than generic mirror star diagonals, degrading the image quality slightly. If you prefer, you can leave the star diagonal out, but this does make viewing through some telescope designs very awkward: you'll need to put the eyepiece and diagonal in to center the telescope on your target then remove them when you plug in the Barlow and the webcam. Removing the star diagonal noticeably changes the focal length of the telescope to such an extent that the defocused image of the planet recorded by the webcam may not be visible on the laptop display. Unless you are very careful, it is very easy to knock the telescope slightly while plugging and unplugging the accessories on the telescope and the focusing it, taking the planet out of the field of view all together. In comparison simply swapping the eyepiece for the webcam doesn't require nearly so much adjustment, and though the image displayed on the laptop will be out of focus, at the very least it will be a big bright blob and so obvious enough to re-center and focus as required. A star diagonal has another useful function in *multiplying the power of a Barlow lens* by about 50 percent if placed before the star diagonal rather than after it as is normal, so a good quality mirror or prism is probably well worth investing in. Good 1.25-inch star diagonals cost around $80 and upwards, and 2-inch ones a little bit more, around $100 for a good, traditionally made one through to over $200 for models with enhanced coatings to minimize the light wasted by scattering on the mirror. However, if all you do is casual stargazing from your back garden or a suburban park, then the difference between a middle of the road model and the more expensive super star diagonals is going to be slight. Where they do become worth considering is if you routinely use high magnifications to observe and image the Moon and planets, or are lucky enough to observe under very dark skies, where the ultimate in light transmission is important, and go hunting for faint deep sky objects. Note also that a 2-inch star diagonal weighs about half a kilo (around a pound) and so may require careful balancing with counterweights if it isn't to put too much strain on the motors and clutches. Models designed for refractors can be fitted to SCTs and Maksutovs using adapters, but taken together these need about 15 cm (6 inches) of clearance within the fork arms of the mount if they aren't to run into the base when pointing towards the zenith. So for example, while the LX 200 telescopes offer enough clearance, the otherwise similar LX 90 doesn't. Instead, look out for "short" 2-inch star diagonals produced especially for SCTs and Maksutovs, for example those from Celestron, Meade and William Optics.

Hardly less important than recording good footage of the planet you're trying to image is accurate tracking. With the Moon tracking is less critical because it is such a large target, in fact it is perfectly possible to sweep across the surface of the Moon with a Dobsonian or an alt-azimuthally mounted refractor simply by pushing the optical tube gently. There isn't much danger of the Moon slipping out of the field of view completely, and if it does, bringing it back into view isn't difficult. In contrast the planets are small: at it's largest Jupiter is little more than one and a half arc-minutes in diameter, far smaller than the thirty arc-minute diameter of the Moon. Moreover, at high magnifications the field of view taken in by a webcam is relatively narrow, too, a typical webcam being used with a ×2 Barlow lens on an f/10 200-mm SCT is only likely to be taking in a field less than fifteen arc-minutes in width. Even very small errors in tracking will be enough to cause perceptible drifting in the position of the planet within the frame, and these make aligning the images afterwards all the more difficult. Check the tracking on your telescope works well beforehand, making sure that the planet stays in the field of view of a high-power eyepiece (around ×200 to ×250 with an f/10 200-mm SCT) for at least several minutes. Getting the polar alignment of an equatorially mounted telescope just right is important, and understanding how an equatorial mount works and testing it out by using the setting circles can pay dividends when it comes to relying on the mount for effective tracking. A stable tripod makes a big difference too, and if your mounting is too light, just focusing the telescope or replacing the eyepiece with a webcam is likely to knock the planet right out of the field of view. Chapter 5 included some ideas for improving lightweight mounts, but if you are serious about astrophotography, a stable, good-quality mount is essential. Go-to telescopes can provide excellent tracking, at least over the short periods required for this sort of photography, but their effectiveness in this regard is dependent on how carefully you aligned it at the start of the evening. If the go-to doesn't work well enough to center the planet in the field of view of a high-power eyepiece, then the tracking won't do at all. A reticule eyepiece is a sound investment if you have trouble judging whether the alignment stars you use are dead center.

Aiming and Focusing the Telescope

Getting the image of the planet dead center on the webcam's CCD is a lot easier said than done. Above all else, properly working magnifying finderscope is essential, and spending a bit of time before hand getting it correctly aligned will go a long way to making your life easier when it comes to get the planet centered on your webcam. Particularly infuriating are the ones with only a single support; these never seem to hold their alignment for very long. Marginally better are finderscopes with two supports but only one of which is adjustable, the other relies on rubber o-ring to hold the finderscope in place. Supposedly, you can adjust the screws on the one support and the o-ring in the other will be flexible enough to allow the finderscope to tilt as required by firm enough to stay aligned once you let go. In practice, these also tend to drift out of alignment, if only because the rubber tends to expand and contract over time and as the weather changes. By far the best design of finderscopes comes with two supports each of which incorporates three adjustable screws. Trying to align these things can still be a maddening experience though, because the arrangement of three screws at 120° to one another isn't as familiar and intuitive to us as four at 90° angles might be. With the main telescope aimed at a distant object you can see with *both* the finderscope and a medium or high-magnification eyepiece at the telescope, the trick is to use these screws to move the finderscope correctly into position. Don't push the finderscope, this achieves nothing at all, as it will do is spring out of alignment when you let go. Making sure that all the screws are tight take two screws on one ring at a time, loosening one and tighten the other, so that the finderscope moves closer to where you want it, and repeat using different pairs of screws as required. What you mustn't do is tighten just a single screw, as all this does is compress the optical tube of the finderscope, and gradually the screw forces finderscope back out of alignment anyway. Zero-power (or unity) finders, normally built around some sort of red or green LED, don't have much use in astrophotography, though they are fine for aligning a go-to telescope. The problem is that the spot of light these things project onto the display window is too coarse to be much use for correctly aligning with the telescope at high magnification.

With the finderscope as your key aid to helping you center the image of a planet in both an eyepiece and on the CCD of your webcam, the next issue is focus. Focusing sounds easy, and on most telescopes is, at least visually. However, with a webcam there is a lag between what you do at the telescope and how it appears on the screen. A Hartman mask is a useful tool for focusing if you find the process too fiddly or suspect that the images you are recording are not quite as sharp as they should be. Essentially a Hartman mask is nothing more than a screen that goes over the aperture of the telescope containing three openings. When the telescope is out of focus, you will see three blobs of light, and only when in focus do they merge to form a single sharp image. You can then remove the mask safe in the knowledge that the telescope is exactly in focus. Some people make their own Hartman masks from cardboard while other prefer to buy them ready made (see the webcam equipment section of Appendix 1 for suppliers), but either way there is no question that these things are very useful indeed.

The focusers on refractors and Newtonian reflectors work by shunting the eyepiece (or webcam) backwards and forwards as necessary, and on the whole these work very smoothly because they hold everything in tight alignment and move along just one axis. However, most catadioptrics work differently, with the eyepiece or webcam remaining in a fixed position while the mirror itself is rotated, moving it backwards and forwards through the optical tube. This way, attaching heavy optical equipment, particularly large cameras and CCDs, doesn't mess up the balance of the telescope or put strain on the eyepiece holder. There is a downside to this though, and that is something called *focus shift*, where the primary mirror wobbles slightly when you focus the telescope, causing the image to move about. At high magnifications this movement can be enough to take the image out of the field of view altogether. Some of the more expensive Maksutovs and SCTs have electronic focusers, which use small motors to wind the focuser that cause fewer vibrations and so reduce the image shift, but more often than not this is something many Maksutov and SCT users feel they have to put up with. This is a shame, as Maksutov and Maksutov Newtonian telescopes in particular are excellent devices for imaging the planets, offering large apertures at low cost but with refractor-like performance. There are after-market solutions, sold primarily for serious astrophotographers using film cameras and CCDs, but if you find image shift a real nuisance with your telescope, they could be worth looking at. Screw-on Crayford focusers from William Optics, JMI and others give your catadioptric a drawtube focuser similar to those on Newtonians and refractors. Depending on the model it will either screw straight into the visual back at the rear of the SCT or Maksutov, or slide into a similar sort of adapter to the ones sold for using refractor star diagonals with these telescopes. There are a couple of catches though. First, none of these add-on focusers is cheap, upwards of $130 for the basic ones and over $200 for the motorized ones. Second, as with 2-inch star diagonals, these things are surprisingly big. Make sure that your SCT or Maksutov has enough clearance between the arms to accommodate the focuser when the optical tube is oriented towards the zenith before buying one of these.

Registering, Stacking and Manipulating the Images

How many frames do you need to make a good final image? You can get surprisingly good results with as few as twenty, but many of the best images use upwards of two hundred, in some case over a thousand. The bottom line is that you want to record as much footage as you can, particularly when the seeing is good. Registering and stacking two hundred frames manually is clearly going to be a tiresome exercise, and most people prefer to use a program like *Registax* or *Keith's Image Stacker* to do the job automatically. These programs differ in the details, but the basic premise is consistent, you choose a nice sharp feature in a *reference frame*, and then the program runs through the other frames and nudges them until they align with this reference frame. Normally, the planetary disc is the chosen feature, but one thing will become obvious if you look at your recorded movies, and that is the way the disc doesn't seem to stay in exactly the same place or hold its shape, instead it seems to bubble and bounce around. This is a result of atmospheric turbulence, and to get around this it is important to only include the better frames from your movies. If you don't do this, the registration process is less effective because the shapes it is trying to match up are less distinct, and consequently stacking the frames is less effective because the details in each frame don't quite line up. It is better to align carefully just twenty sharp frames than to sloppily register a hundred fuzzy ones. While you're looking at your movies, be critical. No amount of processing is going to bring out details that weren't recorded in the first place, what stacking will to is improve the boldness and contrast of the details making them easier to see. If your images are not that good, go back out and try again! All else being equal, there should be some good frames worth aligning, twenty to thirty works fine to begin with, but the more frames you reserve, the better the final image is likely to be.

With luck, stacking will make the individual details more obvious (things like the bands on Jupiter, the Cassini Division in Saturn's rings, or the polar ice caps on Mars are ideal things to concentrate on). Compare your stacked image with a single good frame from the original movie. Can you see these features more clearly? Are contrast boundaries and shadows more obvious? If the answer is no, then stacking didn't help; probably you included some blurry shots among the sharp ones. Try stacking again, this time being even more picky about which frames you include. Does the planetary disc remain sharp edged? If not, something probably went wrong with the alignment procedure and you might want to go back and try again. Once you have the images registered and stacked, it is time to manipulate them as described earlier on in this chapter. Processing planetary images can be very rewarding but the key thing to remember is that *less is more*. It is very easy to apply high levels of unsharp masking or Laplacian filtering with the result that details become exaggerated and artificial looking. Go slowly, nudge

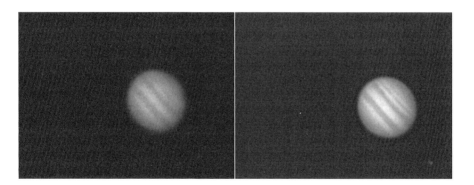

Figure 6.23. You need to stack individual frames (such as the picture of Jupiter on the left) to enhance the detail apparent in the final image (as on the right). Note the improvement in the detail seen on the disc, in particular the equatorial and temperate zone bands the color contrasts between adjacent atmospheric regions. One of Jupiter's moons is also much brighter in the stacked image than in the original.

these filters up to the higher values bit by bit, and check out the image each time before finally applying the chosen filter.

In contrast with the Moon, where there is no particular need to retain color information, color is very important when imaging the planets. Even with a small telescope, the planets are colorful, though perhaps less vividly so than the heavily processed images sent back by space probes and telescopes such as Hubble. Adjacent regions on the surfaces of the planets may differ only modestly, but this contrast is very important to our impressions of these worlds. The equatorial and polar regions of Saturn are all shades of greeny-yellow at the eyepiece but our eyes can easily tell the exact shades apart, and likewise the various albedo features of Mars and the bands on Jupiter. As we've already seen, the *Curves* and *Levels* tools are the ones you need to use to control the shades and tones of the colors used in your images, and both can be used to exaggerate the colors recorded by the webcam and so bring the final image closer to what we see at the eyepiece. The Curves tool in particular is useful for this. There are four options: a combined RGB channel (where all colors are lightened or darkened across the board), and then the separate red, green and blue ones that only change the tones of those primary colors. These will allow you to map a more vivid shade of a color onto the final image compared to the one in the original recorded by the webcam. Just as important can be *Color Balance* (in *Photoshop* within Adjustments section of the Image menu item). Three sliders allow you to shift the color balance between cyan and red, magenta and green, and yellow and blue, respectively. Under some atmospheric conditions, particularly those prevalent in cities, pictures of planets can appear a bit more yellowy than they normally are, and the yellow-blue slider can be used to redress the balance, by pulling the slider towards the blue.

Other Targets: Stars and the Deep Sky

Although the solar system provided by far the best opportunities for enjoying webcam astrophotography, there are others. Stars of various sorts work very well if they are sufficiently bright. A 76-mm (3-inch) refractor will work well enough with stars down to about first magnitude, depending on how dark your skies are, and with a 200-mm (8-inch) reflector third and fourth magnitude stars are possible. Attractively colored stars are of course obvious targets, Antares, Betelgeuse, Vega and Rigel being some of the more obvious stars well known for their brightness as well as their strong coloration. Since the apparent size of stars is even less than that of planets, it is even more important to use a telescope on a steady tripod with efficient tracking and smooth focusing. As with the planets, a certain amount of processing is required for pictures of the stars to look good, but on the plus side registering the frames is pretty straightforward because the stars are small and usually have a nice sharp edge compared to the sky around them. With the brighter zero and first magnitude stars a handful of frames is all that is needed and these can be stacked in a graphics program by hand, and even the fainter stars only need twenty to thirty frames to get a nice looking picture. The

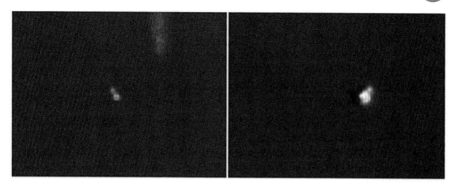

Figure 6.24. Double stars are particularly rewarding targets for webcam astrophotographers because they do not require much processing to look good. Castor (on the left) and Epsilon Bootis (on the right) have been imaged using two different telescope, a small reflector for the first star and a larger SCT for the second. The larger the telescope, the brighter the star and the better the resolution.

colors will need a little work though, because when compared to our visual impressions of these stars at the eyepiece, the colors can seem rather pale. Adjusting the curves is the simplest way to fix this, enhancing the blue or red to get the star looking a bit more impressive. You can also darken the entire background a notch to heighten the contrast using the magic want tool in the same way as described with the Moon: select the sky, play with the tolerance so that it selects just sky and not your star, and then pull down the brightness and step up the contrast.

Double stars are among the most rewarding targets for the webcam astrophotographer. Albireo is easy with a 76-mm (3-inch) refractor or 90-mm (3.5-inch) Maksutov, while Castor and Algeiba are more of challenge but perfectly possible with a 114 to 150-mm (4.5 to 6-inch) Newtonian. A 200-mm (8-inch) telescope lets you take in some classic double stars, including Gamma Andromedae and Epsilon Bootis. Even the famous Double-double in Lyra is possible if you have good optics and great seeing. The Double-double needs *both* magnification and a wide field, so you may need to image each pair separately at high power and then join the two shots together into a mini-mosaic rather as we did with the Moon. With many of these stars, part of the fun of imaging them is in seeing how their contrasting colors come out on the final image. Our eyes see the contrast between a red primary and a blue secondary in rather a subjective way depending on their relative brightness and distance from one another. The color sensitivity of the webcam is different, and you might need to tweak the color balance or the curves to get the two stars looking how you *think* they should!

If you have a large enough telescope (at least 200 mm) and dark enough skies, it is just about possible to record the brightest nebulae with an unmodified webcam. They will be very faint, but stacking can make them more apparent in the final image, but a great deal or processing will be required to increase their overall contrast. The problem with this is that it can also make them look rather artificial. The Orion Nebula is a popular target, being both large and relatively bright. Smaller, but still bright, nebulae like the Ring Nebula in Lyra are possible but as with the stars and planets the need for high magnifications makes them altogether trickier. Modified webcams allow you to cast your net much further because they allow for longer exposures. They can be used to image wide variety of nebulae, as well as open and globular clusters, and even a few galaxies. However, the use of these devices is much closer to the use of astronomical CCD cameras and so falls outside the scope of this book; titles of some good CCD books are give in Appendix 1.

Publishing on the Internet

Taking images to record the things we see is nice enough, but sharing your images is ultimately what astrophotography is all about. After all, the view we see of the Moon or planets when looking into the eyepiece is a personal experience: just you, your telescope, and the cosmos. However, a gallery of your favorite photographs on a web page allows others to understand what it is you go out in the freezing cold at night to see. Some graphics programs include tools for automatically writing the HTML code for web galleries to display your chosen images and the tools for optimizing those images to make them quick to download yet sharp and clear. Some basic design tips are worth stating, though.

Firstly, beauty and simplicity are closely related: the simpler the web page, the better it will look. Resist the temptation to include every animated GIF and musical sound file that you can find on the Web to male your web page seem more interesting. These things distract the user from the actual content, your images, and add significantly to the time it takes for your web page to render on the visitor's browser. Professional graphic designers *never* use these sorts of things, so why should you? Have confidence in the your words and pictures to entertain and inform the visitor. Next, do everything you can to make navigating between pages and accessing the content on your web site easy. Home or index buttons are a great way to help visitors get back to the start if they should get lost. Keep the type legible: brightly colored text against a patterned background is very difficult to read yet widely used. For some reason many of those with an astronomical bent, feel white or yellow text against a backdrop of stars is somehow functional – it's not! If your images are big, put reduced sized thumbnails of them on the gallery page and make them clickable, so that they include a link to the full size image for those who want to see them. If you have a gallery with dozens and dozens of images, then you'll probably want to use thumbnails, and perhaps divide them up into section, one batch for the Moon, the next for Mars, and so on. It is a good idea to include some information with each picture, not just the objects name, but the date you took the image on, what telescope and webcam you used, and any tricks you used to improve the final image. These can be very useful from people who want to learn from you, and sharing your *experience* is just as important as sharing your *results*. Finally, be sure and look at your web page using another computer. The *gamma*, or brightness, of computer screens vary, and photographs that can look clear and bright on one can be very dim on another. Windows and Macintosh computers also use a slightly different screen resolution, and this can make text composed on one computer look very small or too large on the other. Therefore, if you compose your web page on a Mac, you might want to check it looks okay on a Windows PC.

Digital Cameras for Astrophotography

This chapter has concentrated on webcam astrophotography because that is the simplest and least expensive to get good results quickly. However, digital still cameras have become increasingly popular among home users for general picture taking, and they can work surprisingly well for astrophotography. The lenses are removable from the more expensive ones, and consequently these work with traditional SLR camera accessories including telescope adapters, flip mirror systems, and so on. These cameras work with piggyback camera mounts for wide-field imaging, or connected to the optical tube for both prime focus and eyepiece projection photography. Consumer-grade cameras have integral lenses that are not removable. Consequently, these are less flexible, and the only way to use them is via the eyepiece projection method. Some people have had luck simply holding the lens of the camera over the eyepiece and taking a picture of whatever is there, but to get serious you'll need some sort of brace to hold the camera steadily at the eyepiece. Various adapters are available for this purpose, and Appendix 1 includes some sources of these.

One downside with eyepiece projection is vignetting, the tendency for the central portion of the field of view to be brighter than the edges, but the principal shortcoming to digital still cameras as tools for astrophotography is that they take single frames rather than a movie including many frames. This means that in situations where you want to choose a large number of good frames for stacking and processing, as with images of the planets, webcams have the advantage. On the other hand, where a single frame or a small number of them is adequate, as with imaging the Moon or double stars, a webcam is a perfectly serviceable alternative. Digital cameras do have one key advantage, however: some models allow for long exposures (in some cases up to a minute or more thanks to a *bulb* setting similar in use to that on a traditional SLR camera). This makes them much better suited to deep sky imaging than unmodified webcams. The better ones have built-in noise reduction that further improves the quality of long exposure images. The Nikon Coolpix 4500 is one such camera, and very popular with amateur astrophotographers, and capable of not just nice images of the Moon and planets, but of globular clusters and other challenging deep sky objects. Other digital cameras don't offer long exposures, and these are limited to shots of the Moon and the brighter planets.

In operation, use a digital camera in much the same way as a webcam, except this time leave the eyepiece on the telescope, and use the adapter to place the camera over the eyepiece. A low to medium magnification eyepiece are best to begin with, although this will depend somewhat on your telescope; 32 to 18-mm

Plössls are particularly popular and a range of suitable adapters are produced by Scopetronix and other manufacturers. Focus the camera on infinity and use a wide-angle setting if it has a built-in zoom. To begin with aim the camera at something bright, perhaps the Moon, or a planet; otherwise, a bright star will do. Then zoom into the object using the camera but focus carefully using the telescope. What you are trying to do is get the biggest image possible with the camera's lens (so that it covers the greatest proportion of the CCD inside the device) while using the telescope to take care of the focusing. Once you are happy you have everything working properly, return the camera to its wide field view and then move the telescope to the target, such as a globular cluster. Then return the zoom to your chosen narrow field setting, and then take the long-exposure picture. With luck, you should have captured a nice shot of the target. This approach takes a little practice and does rather depend on your setting circles or go-to alignment being good to begin with, but once you have the hang of it works remarkably well. A single long-exposure image might work for some objects, but often a number of stacked images will work even better. This is particularly true with nebulae, which are often relatively low contrast objects. If a few images (at most a few dozen) are stacked manually in a graphics program, surprisingly good pictures are possible, even with a trace of color!

Moving on to Film and CCD

Traditional film and CCD astrophotography (other than with webcams and digital cameras) falls outside the scope of this book, but both provide plenty of opportunities for integration into the digital lifestyle. If you scan in prints taken using a traditional camera, applications such as Photoshop allow manipulations of contrast and color balance, and may be used the same way with the pictures taken with image files created by CCD cameras. The key difference with traditional film and CCD cameras is that both offer a range of long-exposure settings, and so image fainter objects more effectively. Film cameras are very versatile, and can be used for simple wide-field views of the stars when riding piggyback on an equatorially mounted telescope, through to prime focus and eyepiece projection photography of everything from galaxies to the Moon. CCD cameras are much more expensive than even good film cameras, but have the advantage of being able to do much the same thing but with shorter exposure lengths, and of course eliminate the need for buying and developing film. As noted earlier in this book, the majority of CCD cameras require a Windows PC to operate, though third-party software can allow some models to work with Macintosh and Linux computers. A more serious downside to CCD imaging is the cost, since even the basic models cost upwards of the thousand dollars and the better ones several times that. Therefore, while they offer the amateur the chance to take the sorts of pictures previously only possible with professional equipment, CCDs remain a minority pursuit within the hobby generally. If the price tag doesn't put you off them, and you fancy learning more, Appendix 1 includes the titles of some useful books for those interested in moving into film or CCD astrophotography.

APPENDIX 1

Reading and Resources

Further Reading

Telescopes and Astronomical Equipment

The following are books that offer particularly sage advice on selecting and using astronomical equipment. While many astronomy books include short accounts of the basic types of telescope and useful type of accessories to look out for, the authors of these books name the good and shame the bad, which makes them invaluable companions when shopping.

- *Choosing and Using the Schmidt–Cassegrain Telescope.* R. Mollise, Springer-Verlag, 2002, London, UK.
- *How to Use a Computerized Telescope: Practical Amateur Astronomy Volume 1.* M. A. Covington, Cambridge University Press, 2002, Cambridge, UK.
- *Star Ware (3rd Edition).* P. S. Harrington, John Wiley & Sons, 2002, USA.
- *Stargazing with a Telescope.* R. Scagell, Philips, 2000, London, UK.
- *The Backyard Astronomer's Guide (2nd Edition).* T. Dickinson and A. Dyer, Firefly Books, 2002, USA.
- *Using the Meade ETX.* M. Weasner, Springer-Verlag, 2002, London, UK.

Star-Hopping Techniques

Star hopping is a great way to combine your planetarium program with your telescope; use them to produce charts that are either viewed on the computer screen or printed off to use away from the computer. Here are two of great books for learning star- hopping techniques and expanding the range of things you observe each evening.

- *Star-Hopping: Your Visa to Viewing the Universe*. R. Garfinkle, Cambridge University Press, 1997, UK.
- *Turn Left at Orion*. G. Consolmagno and D. Davis, Cambridge University Press, 2000, UK.

Catalogs of Deep Sky Objects

If you want to put together an observing program based on a certain type of object, you'll need a source book to get ideas from. The two star hopping books mentioned above are a good place to start, and the following will provide thousands more objects for more experienced observers.

- *Burnham's Celestial Handbook*. R. Burnham, Dover Books, 1978, USA.
- *Field Guide to the Deep Sky Objects*. M. Inglis, Springer-Verlag, 2001, London, UK.

CCD and Advanced Imaging Methods

The details and methods involved in advanced electronic imaging using specially designed astronomical CCD cameras instead of webcams and digital cameras falls outside the scope of this book. Fortunately, a good number of books already exist on the topic, to which the reader is referred.

- *CCD Astronomy: Construction and Use of an Astronomical CCD Camera*. C. Buil, 1991, Willman–Bell, USA.
- *Choosing And Using A CCD Camera*. R. Berry, Cambridge University Press, 1992, UK.
- *Handbook of Astronomical Image Processing (includes Astronomical Image Processing for Windows)*. R. Berry and J. Burnell, Willman–Bell, 2001, USA.
- *Practical Astrophotography*. J. R. Charles, Springer-Verlag, 2000, London, UK.
- *A Practical Guide to CCD Astronomy*. P. Martinez and A. Klotz, 1997, Cambridge University Press, UK.

Charting and Utility Software

Planetarium and Moon-Mapping Software

Star charting software is among the most useful available to astronomers, and there are applications available for all budgets and levels of observing skill. Such applications can be used either to produce tailor-made star charts for printing, or in the field alongside the telescope, in which case some sort of night vision mode is essential. Moon-mapping software produces much more detailed charts of the lunar surface, allowing the user to identify craters and other features much more easily than with traditional books and maps.

- *2sky* (http://in2space.com) for Palm OS. Basic version includes stars down to 6th magnitude plus five hundred deep sky objects. Additional databases can be purchased for ten dollars a throw, including the full NGC/IC catalogs. Commercial.

- *Alpha Centaure* (http://astrosurf.com/alphacentaure/english/index1.htm) for Windows. Lightweight, easy to use planetarium including a good selection of bright stars, Messier and NGC objects, comets and so on. Includes a night vision mode and a user-configurable sunspot plotter. Freeware.
- *Cartes du Ciel* (http://www.stargazing.net/astropc/index.html) for Windows. Full-featured, easy to use planetarium program. Freeware.
- *Deep Sky 2003* (http://www.deepsky2000.com) for Windows. Big, comprehensive planetarium with hundreds of thousands of deeps sky objects and millions of stars and advanced features like telescope control and CCD image processing. Commercial.
- *Equinox* (http://www.microprojects.ca) for the Mac. Impressively appointed but low-cost planetarium program including such luxuries as go-to telescope control and observing lists. Shareware.
- *KStars* (http://edu.kde.org/kstars) for Linux. Powerful, Internet-savvy open source planetarium, and probably the easiest to use for Linux or UNIX generally. Can be modified to run on the Mac as well. Freeware.
- *Lunar Map Pro* (http://www.riti.com) for Windows. Powerful lunar atlas for serious observers. Commercial.
- *MegaStar* (http://www.willbell.com) for Windows. Widely considered to be the best planetarium program for advanced deep sky observers. Includes huge catalogs of deep sky objects, support for most go-to telescopes and mounts, CCD camera utilities and more. Commercial.
- *Palm Planetarium* (http://www.aho.ch/pilotplanets) for Palm OS. Includes features such as telescope control (via the serial port) and a night vision mode (on color systems). Shareware.
- *RedShift* (http://www.maris.com) for Windows and Mac. Primarily devised as an educational tool but includes a planetarium as well. Commercial.
- *SkyChart* (http://skychart.sourceforge.net) for Windows and Linux. Still being developed and based on the popular Windows application *Cartes du Ciel*, this is currently a usable if simple planetarium program. Freeware.
- *SkyChart 3* (http:www.southernstars.com) for Windows and Mac. Low cost but sophisticated planetarium program. Sky Sight, a Mac CCD program, is also available for free.
- *SkyMap Pro* (http:www.skymap.com) for Windows. Heavyweight program for advanced amateurs containing just about every feature imaginable. Commercial.
- *Stargazer's Delight* (http://www.stargazersoft.com) for the Mac. Simple and fun lightweight planetarium including some neat tutorials and animations, as well as useful tools such as plots of Jupiter's moons. Shareware.
- *Starry Night* (http://www.starrynight.com) for Windows and Mac. Photorealistic planetarium program available in basic *Backyard* and advanced *Pro* versions. The *Pro* version includes much larger object catalogs, telescope control and a planning tool for creating and optimizing observing programs. Commercial.
- *Stellarium* (http://stellarium.free.fr) for Windows, Mac and Linux. Beautiful if somewhat limited "eye-candy" planetarium offering some of the best virtual stargazing around. Freeware.
- *The Digital Universe* (http://www.syz.com) for Windows, Mac and Amiga. Intuitive and attractive planetarium program including a very fine astronomical encyclopaedia with hundreds of pictures and long, well-research and very detailed articles. Comes bundled with a 3-D star simulation called *3DStars* and red/blue spectacles for getting the full effect with. Commercial.
- *TheSky* (http://www.bisque.com/) for Windows, Mac and Windows CE. Sophisticated and easy to use planetarium with many useful features. Available in various editions,

from the basic *Student* edition through to the powerful *Level IV* version that including telescope control, image manipulation and editing, and large deep sky object catalogs. Commercial.

- *Virtual Moon Atlas* (http://astrosurf.com/avl/UK_index.html) for Windows. Useful and easy to use lunar atlas. Freeware.
- *Voyager* (http://www.carinasoft.com) for the Mac. Full-featured planetarium program although lacking integrated go-to telescope control. A lightweight version of this application for Windows and Mac is also available, called *SkyGazer*. Commercial.
- *XEphem* (http://www.clearskyinstitute.com) for Windows, Mac and Linux. Very powerful astronomical ephemeris including a planetarium mode as well as maps of Mars and the Moon, satellite imagery of the Earth, FITS image views and more. CSI also produce a star-charting program for the Sharp Zaurus series of palmtop computers. Commercial.

Telescope Control, Logging and Utility Software

Various utilities exist to improve the lot of the amateur astronomer, including electronic guidebooks to the night sky, software for creating observing lists and stepping go-to telescopes through them and utilities for changing the colors on a laptop to a more night-vision friendly mode.

- *AstroPlanner* (http://www.ilangainc.com/astroplanner) for Windows and Mac. This is a multi-purpose tool that includes deep sky observing list generation, calculations of the "best pair" of stars to use for aligning go-to telescopes, and go-to telescopes control (via a serial cable). Shareware.
- *David Paul Green's Free Software* (http://www.davidpaulgreen.com/software.html) for Windows and Mac. A great suite of tools for logging observations, including ones for the Messier and Caldwell lists. Freeware.
- *Night Vision* (http://www.adpartnership.net/NightVision/index.html) for Windows and Mac. A night vision utility that darkens and tints the screen allowing a laptop to be used in the field without ruining dark adaptation. Freeware.
- *NightMaster* (http://www.ilangainc.com/nightmaster) for the Mac. Another night vision utility. Includes red, green and blue sliders that allow some very funky display configurations! Freeware.
- *Scope Driver* (http://www.adpartnership.net/ScopeDriver) for Windows and Mac. Fast, lean and easy to use list-based observing utility. Shareware.

X Windows and Professional Astronomical Software

Many serious astronomical programs are available for download and use from the Internet, and though of limited practical value, can be fun and educational. In some cases they will run in the traditional Windows or Mac operating system but often they require the presence of an X Windows server of some sort.

- *Apple X11* (http://www.apple.com/macosx/x11) for the Mac. The simplest X Windows server to install on a modern Mac computer. Runs UNIX software alongside traditional Mac programs. Freeware.
- *DS9* (http://hea-www.harvard.edu/rd/ds9) for X Windows. An image analysis program designed for use by professional astronomers, but amateurs will find the ability to

view FITS files (as well as a variety of other graphics file formats) useful as well. Freeware.

- *Cygwin* (http://cygwin.com) for Windows. A popular Linux emulator, not as easy to install as the commercial alternatives, but effective nonetheless. Freeware.
- *Fink* (http://fink.sourceforge.net) for Mac. Fink is a project consisting of ported versions of UNIX applications adapted to run on X Windows on Mac hardware, including versions of *KStars, StarPlot and Nightfall.*
- *GeoVirgil* (http://www.siliconspaceships.com) for Windows and Mac. A Java-based application that accesses and displays NASA images of Mars, Venus and other solar system bodies. A companion program called AstroVirgil does the same thing for x-ray images from the Chandra space telescope. Freeware.
- *Nightfall* (http://www.lsw.uni-heidelberg.de/~rwichman/Nightfall.html) for X Windows. The program simulates the orbits of binary stars and shows things like the way the shape of each star distorts under the gravitational influence of its companion. Freeware.
- *OroborOSX* (http://oroborosx.sourceforge.net) for the Mac. Augments *XFree86* by integrating it more completely with the Mac operating system, for example adding the ability to copy and paste between programs. Freeware.
- *Partiview* (http://www.haydenplanetarium.org/hp/vo/du/index.html) for Windows, Mac and UNIX. A three-dimensional atlas of the Milky Way and beyond; very pretty and ideal for educators as well as curious amateur astronomers. Freeware.
- *SETI@Home* (http://setiathome.ssl.berkeley.edu) for Windows, Mac and UNIX. Various applications that allow computers with a connection to the Internet to download and process SETI data. Freeware.
- *StarPlot* (http://starplot.org/index.html) for X Windows. Simple and attractive stellar cartography program designed to show the relative positions of stars to one another in three dimensions. Freeware.
- *Virtual PC* (http://www.microsoft.com/windowsxp/virtualpc) for Windows and Mac. Because it emulates the hardware that can run a Linux operating system, it is relatively straightforward to install and run a full-blown Linux operating system onto Virtual PC instead of some version of Microsoft Windows. Commercial.
- *WinaXe* (http://labf.com/index.html) for Windows. A sophisticated but relatively easy to install and use X Windows emulation package that installs onto computers from Windows 95 to XP. Commercial.
- *XFree86* (http://mrcla.com/XonX) for the Mac. A basic X Windows server that runs UNIX programs but lacks some of the niceties of the usual Macintosh front end; these can be added using *OroborOSX*. Runs UNIX software ported to the Mac. Freeware.

Go-To Telescopes and Accessories

All-in-One Go-To Telescope Manufacturers

Only two companies mass-produce all-in-one go-to telescopes, and which is the better of the two is a popular topic for discussion among astronomy hobbyists! Catadioptric telescopes of one sort or another dominate the computerized telescope ranges of both companies.

- *Celestron* (http://www.celestron.com). Major manufacturer of consumer level telescopes and accessories. NexStar series of all-in-one go-to telescopes include refractor, reflector and catadioptric designs. Links to worldwide dealers.

- *Meade* (http://www.meade.com). Major manufacturer of consumer level telescopes and accessories. Autostar series of all-in-one go-to telescopes include refractor, reflector and catadioptric designs. Links to worldwide dealers.

Go-To Mounts and Mount Upgrades

Go-to mounts offer much more flexibility in terms of what optical tube assembly is used, which is ideal if you want to use a high-quality apochromatic refractor or short focal length Newtonian sold as an optical tube alone.

- *Alpine Astro* (http://www.alpineastro.com). US distributor of Baader Planetarium products, including the adapters for fixing optical tubes to the smaller Celestron go-to telescope mounts.
- *Astro-Physics* (http://www.astro-physics.com). Manufacturer of top-quality telescopes and mounts, including the GTO series of go-to mounts. Links to worldwide dealers.
- *Baader Planetarium* (http://www.baader-planetarium.de). Manufacturer and distributor of various astronomical accessories including a repackaged NexStar go-to mount suitable for use with small refractors and catadioptric telescopes, including Leica and Zeiss spotter scopes. In German; some online ordering available, plus links to worldwide dealers.
- *Jim's Mobile* (http://jimsmobile.com). Manufacturer and distributor of a wide range of useful accessories for astronomers, including the NGC-MAX DSC system and flight cases for go-to telescopes.
- *Losmandy* (http://www.losmandy.com). Manufacturers of high-quality telescope tripods and mounts including a go-to system known as Gemini. Losmandy also produce the mounting rings and counterweights needed to attach small telescopes to larger ones. Online ordering available.
- *Lumicon* (http://www.lumicon.com). A division of Parks International, the Lumicon range includes the Sky Vector DSC system as well as various optical and light pollution filters, illuminated reticule eyepieces and astrophotography equipment. Online ordering available.
- *Orion* (http://www.oriontelescope.com). Major distributor of mass-market astronomical equipment in the US, including the IntelliScope DSC system for their XT-series of Dobsonian telescopes. Online ordering available.
- *Sky Engineering* (http://skyeng.com). Manufacturer of DSC systems for equatorially mounted and Dobsonian telescopes.
- *StarMaster* (http://www.starmastertelescopes.com). Top-quality Dobsonian manufacturer. One upgrade available is the Sky Tracker go-to system.
- *Takahashi* (http://www.takahashiamerica.com). Manufacturers of very high-quality Japanese-made refracting telescopes and mounts, including a go-to system known as Temma II. Links to US dealers.
- *Tech2000* (http://homepages.accnorwalk.com/tddi/tech2000). Manufacturer of various astronomical accessories including motorization kits compatible with Dobsonians and DSC systems. Online ordering available.
- *Vixen* (http://www.vixen-global.com). Japanese manufacturer of good quality telescopes and mounts, plus accessories such as the versatile SkySensor 2000 go-to upgrade package for their equatorial mounts. Links to worldwide dealers.

Manufacturers and Distributors of Useful Accessories

The following suppliers provide equipment that can be used for expanding a go-to telescope, as described in this book.

- *Apogee* (http://www.apogeeinc.com). Distributors of various telescope accessories including replacement star diagonals and focusers for SCT and Maksutov telescopes, camera adapters and collimators. Online ordering available.

- *Broadhurst, Clarkson and Fuller* (http://www.telescopehouse.co.uk). Distributors of their own line of Meade ETX and LX 200 series telescope accessories, as well as others, though primarily Meade and Tele Vue.

- *Kendrick Astro Instruments* (http://www.kendrick-ai.com). Distributors of their own line of accessories for telescopes as well as others, including computerized mounts from Losmandy, Takahashi, Tele Vue and Vixen. Online ordering available.

- *ScopeStuff* (http://www.scopestuff.com). Manufacturers of telescope various accessories including mounting rings and counterweights for attaching small telescopes and cameras to large go-to telescopes such 200-mm (8 inch) SCTs. Online ordering available.

- *ScopeTronix* (http://www.scopetronix.com). Distributors of a wide variety of ETX, LX and NexStar upgrades and accessories, including tripods, counterweights, solar filters, external battery packs and replacement finders. Online ordering available.

- *SkyPointer* (http://www.skypointer.net). Manufacturers of a pen-sized laser pointer that can also be used as a finder device. Online ordering available.

Go-To Telescope Support and Commentaries

There are several web sites devoted to go-to telescopes of various sorts, some of which have become real focal points for mutual support and discussion between amateurs using specific designs of instrument.

- *Jan's LX 90 Pages* (http://m1.aol.com/kewtasheck/lx90.html). A rich seam of information on using and expanding the LX 90 SCT, including sections on updating the handset and improving pointing accuracy.

- *LXD55.com* (http://www.lxd55.com). Reviews, upgrades, astrophotography and more make this an essential read for owners of Meade's LXD 55 family of reflectors and refractors.

- *Meade Advanced Products Users Group* (http://www.mapug.com). Eclectic collection of opinions and e-mails rather than articles, this web site offers some useful information for owners of ETX and LX series telescopes.

- *NexStar Resource Site* (http://www.nexstarsite.com/NUG.htm). Arranged as a companion to a forthcoming book covering all aspects of NexStar use, this site includes many useful articles and downloads.

- *NexStar Web site* (http://home.att.net/~nexstar/index.html). Not regularly updated but still useful site with reviews and commentaries on the first generation of NexStar telescopes.

- *Weasner's ETX Home Page* (http://www.weasner.com). Exceptionally valuable resource for Meade ETX telescope users covering every imaginable aspect of their use from troubleshooting go-to reliability through to astrophotography.

Webcam and Digital Astrophotography

Cameras and Adapters

Webcams can be used as they are out of the box, or modified to extend the exposure lengths and reduce the noise apparent on the images. The following links include web sites detailing their use and modification or sell useful accessories.

- *Ash's Astronomy Pages* (http://astro.ai-software.com). Descriptions of various webcam modifications together with some nice galleries, including a Messier object album!
- *Astrocam.org* (http://www.astrocam.org). Webcam modifications, galleries and help files; very useful web site.
- *Kendrick Astro Instruments* (http://www.kendrick-ai.com). Distributors and manufacturers of various telescope accessories including the "Kwik Focus" Hartman mask. Online ordering available.
- *Long exposure webcams* (http://home.clara.net/smunch/wintro.htm). Includes schematics and explanations of various builds and comments on which cameras they work with.
- *QUCAIG* (http://www.qcuiag.co.uk). QUCAIG is short for the Quick Cam and Unconventional Imaging Astronomy Group, and is a useful site for image processing and descriptions of webcam modifications.
- *SAC Imaging* (http://www.sac-imaging.com/main.html). Manufacturer and distributor of low-cost CCD cameras, a significant step up from a regular webcam in performance and a fine alternative to a do-it-yourself webcam modification. Online ordering available.
- *Sarawak Skies* (http://www.angelfire.com/space2/tgtan). Nice tutorials on image processing, the fundamental step to getting satisfying images from webcam movies.
- *ScopeTronix* (http://www.scopetronix.com). Besides the telescope accessories mentioned above, ScopeTronix also produce adapters for connecting a large number of digital cameras and camcorders to telescopes. Online ordering available.
- *Steven Mogg's Webcam Adapters* (http://webcaddy.com.au/astro/adapter.htm). These adapters provide an easy way to connect a webcam to a telescope, and the range includes adapters for most webcam models. Adapters also available for SLR and digital cameras. Online ordering available.
- *William Optics* (http://www.william-optics.com). A wide variety of adapters for fitting Fuji, Nikon, Sony and other popular digital cameras to telescopes. Online ordering available.

Astrophotography and Image Processing

Webcam astrophotography has become very popular, not least of all because the cameras are inexpensive and compatible with most computers and telescopes. Registering, stacking and processing webcam images does require specialized software though, but fortunately most of this is available on the Internet for free.

- *Adobe Photoshop* (http://www.adobe.com) for Windows and Mac. Powerful, surprisingly easy to use image editing software. Good, but expensive. Commercial.

- *AstroStack* (http://www.astrostack.com) for Windows. Very powerful and rightly popular application for combining frames from a webcam movie into a single high-resolution image. Shareware.

- *AstroYacker* (http://home.iprimus.com.au/rodkennedy/Astro/Jerra.html) for Mac. AstroYacker manipulates webcam movies prior to stacking, for example rotating frames to compensate for field rotation so that they align better.

- *The GIMP* (http://www.gimp.org) for Windows, Mac and Linux. Open-source image editing software; a great zero-cost alternative to *Adobe Photoshop*. Freeware.

- *GIMP for Windows* (http://www.gimp.org/~tml/gimp/win32/) for Windows. Windows-native port of *The GIMP*. Freeware.

- *JImage* (http://rsb.info.nih.gov/ij) for Windows, Mac and Linux. Java-based image processing application. Fast, relatively easy to use and comes with built-in image stacking tools. Freeware.

- *Jasc Software* (http://www.jasc.com) for Windows. A popular consumer-level application that accomplishes many of the things possible with *Adobe Photoshop* but at a significantly lower cost. Commercial.

- *Keith's Image Stacker* (http://www.unm.edu/~keithw/software.html) for the Mac. Broadly equivalent to *AstroStack*. Shareware.

- *Macam* (http://webcam-osx.sourceforge.net) for the Mac. Image capturing software for USB webcams including many designs that are not otherwise Mac compatible. Freeware.

- *Photoshop for Astrophotographers* (http://www.astropix.com/pfa/pfa.htm) for Windows and Mac, an electronic book on a CD including tips and tutorials for using this application with both traditional film and electronic images. Commercial.

- *Qastrocam* (http://3demi.net/astro/qastrocam) for Linux. Source code for building webcam capturing and stacking software for Linux computers. Compatible with a wide variety of webcam models. Freeware.

- *Registax* (http://aberrator.astronomy.net/registax/index.html) for Windows. Sophisticated software for aligning frames from webcam movies precisely and including a great many tools for processing the resulting images. Freeware.

Web Sites: Reviews and Resources

Amateur Astronomical Societies

- *American Association of Amateur Astronomers* (http://www.corvus.com/index.html). A very rich resource with articles on all manner of topics, from quantifying seeing conditions through to detailed descriptions of the constellations. Also plenty of links to other resources, an astronomy store, and of course membership information. Highly recommended even for amateurs outside the US.

- *Astronomical Society of Australia* (http://www.atnf.csiro.au/asa_www/astro.html). The web address given here isn't to the main site of the ASA, which is primarily a professional body, but to their comprehensive list of amateur astronomy societies in the southern hemisphere.

- *British Astronomical Association* (http://www.britastro.org/main/index.html). Home page of the BAA, which is divided up into various observing sections covering topics like Saturn and Aurorae.

Astronomy Resources and Information

- *Clear Sky Clock* (http://cleardarksky.com/csk). Home page for the Clear Sky Clocks used by many amateurs to predict observing conditions for a few nights in advance. The data comes from the Canadian Meteorological Centre, and there are links on this page to clocks that cover most of North America. Very useful.

- *Digitized Sky Survey* (http://www-gsss.stsci.edu/DSS/dss_home.htm). Put together by the Space Telescope Science Institute using the Oschin Schmidt Telescope on Mt. Palomar and the UK Schmidt Telescope in New South Wales, Australia. Several programs (such as *XEphem*) allow users to access these images easily.

- *Ken's Telescope Calculator* (http://www.klhess.com/telecalc.html). Determines aspects of telescope performance such as magnification as well as more tricky measurements like true field and exit pupil using JavaScript. Easy to use, and very useful, and comes complete with some of the most popular eyepieces built-in!

- *FAQ About Collimating a Newtonian Telescope* (http://zebu.uoregon.edu/~mbartels/ kolli/kolli.html). Another good collimation site, including information on the available tools, such as laser collimators.

- *Seeing Forecast for Astronomical Purposes* (http://www.cmc.ec.gc.ca/cmc/htmls/ seeing_e.html). A useful explanation of the five-point seeing scale used by many amateur astronomers and by online "Clear Sky Clocks". Complete with a very useful animation that shows what a star under each point along the seeing scale would look like.

- *Sky & Telescope: Saving Dark Skies* (http://skyandtelescope.com/resources/darksky). Includes discussions of light pollution and how to minimize its effect, plus a detailed explanation of the Bortle Dark Sky scale.

- *Sky Transparency Forecast for Astronomical Purposes* (http://www.cmc.ec.gc.ca/cmc/ htmls/transparence_e.html). As above, for transparency.

- *Thierry Legault's "The Collimation" page* (http://perso.club-internet.fr/legault/ collim.html). A very detailed and helpful site describing what collimation is, how it affects astronomical images, and the ways to correctly collimate a telescope. Essential reading for owners and users of reflecting telescopes.

Equipment Reviews

Astronomical equipment reviews are extremely popular, and many amateurs put up notes on their equipment up on their web sites to share with others. Review sites like the ones listed below take this further in one of two ways: either *compiled* from reviews undertaken by a single author, or *edited* from submitted reviews written by many different authors. Single-author web sites are more consistent in quality and methods, making the comparisons between telescopes and accessories more meaningful, but the workload on one person does mean the site expands only slowly. On the other hand multi-author sites are much more dependent on the quality of the submissions made to them; at their best, with a peer-review process, a constant level of quality can be maintained as well as a much broader and faster evolving range of reviews. The following include some of the most popular and respected review sites.

- *Affordable Astronomical Equipment Reviews* (http://members.tripod.com/irwincur) edited by Curt Irwin. A review sites focused on low-cost astronomical equipment, such as telescopes costing $1500 or less. The quality of the reviews is variable, but there is a good range and plenty of useful information. There is also a spin-off mailing list that complements the site giving amateurs on a budget a forum for discussing equipment and techniques.

- *AppleLust* (http://www.applelust.com/scitech) edited by David Schultz. The science and technology section of AppleLust includes the largest collection astronomy program reviews on the Internet, primarily for the Mac OS but including some for Linux and Windows.

- *Cloudy Nights Telescope Reviews* (http://www.cloudynights.com) edited by Allister St. Claire. One of the best multi-author sites and the only one with a formal peer-review process. A team of experts reads the reviews first, and then they discuss their conclusions with the author. This works to improve consistency and objectivity, and helps authors to elucidate their observations and opinions more clearly.

- *Excelsis* (http://www.excelsis.com) edited by Excelsis Consulting. The great variations in the quality of the reviews at this site is balanced by the enormous breadth they cover, and this site is a useful first stop for information on a wide range of topics including telescopes, eyepieces, retailers and books. Although visitors are able to submit reviews freely, a form of weighting does exist to mark out good reviews from unreliable and poorly written reviews ones.

- *Heretics Guide to Choosing and Buying Your First Telescope* (http://www.findascope. com) compiled by Michael Edelman. Very detailed explanation of the factors to consider before purchasing a telescope. Somewhat partial and idiosyncratic, but useful nevertheless.

- *Scope Reviews* (http://www.scopereviews.com) compiled by Ed Ting. This is one of the oldest and most respected review sites and probably the one against which all the others are compared. Ed Ting's reviews are objective and balanced, taking into account factors like price and ease of use as well as optical quality, and cover a good range of equipment from small aperture reflectors through to top of the range apochromatic refractors. *Scope Reviews* is divided up into sections some of which are logical enough (e.g., reviews of all the Radian eyepieces) but others are simply chronological admixtures of whatever was being reviewed at that time.

- *Todd Gross' Weather and Astronomy Site* (http://www.weatherman.com) compiled by Todd Gross. Like *Scope Reviews* this is a single author site and as such there are logical and detailed comparisons between the various telescopes and accessories discussed. The various reviews are generally balanced and well written, and although not laid out as clearly as some, this is still one of the best review sites out there.

Observing

There are many different web sites devoted to particular aspects of astronomical observing on the Internet, and only a few can be listed here. Some are straightforward lists of things like double stars or deep sky objects, but more interesting perhaps are those web sites that combine text, images, sound and video.

- *33 Doubles* (http://www.carbonar.es/s33/33.html). Observing projects and commentaries based around double and multiple stars. Aimed at amateurs using equipment ranging from binoculars through to large aperture telescopes.

- *Antonio Cidadão's Home Page* (http://astrosurf.com/cidadao). Truly spectacular images of the Moon and planets combined with detailed notes and diagrams make this web site an essential stop for those interested in solar system observing.

- *Hitchhiker's Guide to the Moon* (http://shallowsky.com/moon). A reference guide for lunar observers by Akkana Peck and others. The prime attraction is a map of the Moon that is illuminated according to its phase and divided up into small sections. Click on any one of these to find out about interesting features in that region. Based on the highly praised but out-of-print guidebook to the Moon by Anton Rükl.

- *Inconstant Moon* (http://www.inconstantmoon.com). An unusual multi-media astronomical web site that nicely shows off the potential of the Internet for mixing different sorts of educational and entertainment approaches. Moon-themed music plays in the background (this can be switched off easily enough) while the visitor uses interactive tools like a lunar calendar and an illustrated, hyperlinked encyclopaedia of lunar features. A selection of the sights best seen on the Moon that day is offered together with lists of events such as lunar eclipses.

- *Invitation to the Moon* (http://mo.atz.jp/index-e.htm). Morio Higashida's web site describes many of the most interesting features on the Moon and is particularly inspirational for those observers looking for ideas of things to image using a webcam or digital camera.

- *Observing With A 6″ Reflector* (http://www.geocities.com/the_150mm_reflector/). Alistair Thomson's web site includes observing ideas and reports submitted by amateurs using a wide variety of telescopes, not just the 6-inch reflector suggested by the title.

- *Royal Astronomical Society of Canada's Finest NGC observing list* (http://www.seds.org/messier/xtra/similar/rasc-ngc.html). Although the deep sky objects that make up the NGC tend to be overlooked by beginners in favor of the Messier Catalog, there are some nice objects hidden among the hundreds of faint and commonly rather unimpressive entries. This page lists many of them, and has them arranged by season to help you see them at their best.

Seeing, Transparency and Darkness

Estimating Seeing Conditions

Many beginners find it difficult to judge seeing objectively. Basically the seeing depends on thermal currents and other movements in the air. By looking at a star at high magnification (30–50 times per inch of aperture) and then taking the image out of focus, it is really quite easily to spot these air currents. Under good seeing conditions the image, though now blurred, will be steady and the Airy disc and the diffraction rings will be clear, but if the seeing is bad the Airy disc and diffraction rings will be rapidly scintillating and difficult to distinguish. Of course this assumes the telescope is properly collimated and that it has been allowed time to cool down. In fact warm air currents inside a telescope that has just been brought outside essentially mimic bad seeing, and that is why telescopes need to reach thermal equilibrium with the night air before they can deliver good images. Many astronomers simply talk about "poor" or "good" seeing to describe the conditions they find themselves observing under, but more neatly divided scales do exist. The American Association of Amateur Astronomers (AAAA) and the popular online Clear Sky Clocks use a five-point scale based on that developed by a planetary astronomer by the name of E. M. Antoniadi, as follows:

(I)   Star image appears to be a "boiling" blob with no differentiation between the Airy disc and diffraction rings. This is the poorest seeing, described as "severely disturbed" by the AAAA, and even at low power stars looks blurry. No detail on planets, and views of the Moon are disappointing and difficult to focus as the craters and seas seem to bubble and move.

(II) Star still boiling and difficult to focus, but at low powers at least images are acceptable if not sharp. Some differentiation between the brighter central part of the star and the fainter edge, but with no sign of the Airy disc or diffraction rings as such. The AAAA calls this "poor seeing".

(III) Star now divides clearly into the Airy disc and the diffraction rings, although the rings are more incomplete arcs that complete rings, and will seem to move about a bit under the influence of the slight air currents. The Airy disc will be approximately circular, though again there may be constant changes to its shape as well. This is the sort of night where it becomes worth staying out and looking at the planets and Moon with a reasonable expectation of seeing some detail at moderate powers, what the AAAA calls a night of "good seeing".

(IV) Star with clear and sharp Airy disc and diffraction rings. The central disc is uniformly bright and fairly steady, and the rings will be more or less complete with only small gaps here and there. The AAAA calls this "excellent seeing" and you can expect good images of the Moon, planets and double stars at moderate powers, though at high magnifications detail, though evident, will appear a little blurry.

(V) Star displays textbook-quality Airy disc and diffraction rings: the disc is circular and steady, and the rings around it are prefect circles. Described by the AAAA as "perfect seeing" this is the sort of night amateur astronomers dream about. Magnification can be ramped up to the theoretical limits of the instrument without noticeable image breakdown, and views of all objects will be steady, crisp and detailed.

Estimating Transparency

By day at least it is easy enough to appraise transparency; the clearer the sky of clouds, and the deeper the shade of blue the sky is, the better the transparency. On really good days the sky will be a deep blue right above your head, almost violet. These promise dark nights when you hope the Moon isn't above the horizon and you can get some real deep sky observing done. Transparency is how much light gets from the object being viewed to the observer without being scattered by haze, dust, pollen and other things in the air. Although many beginners confuse transparency with seeing, the two things are quite different, and in fact they aren't usually good at the same time: steady seeing tends to be associated with low transparency, and vice versa. The reason for this is the clear, cold air that is most transparent is also the most mobile, and will, for example, be agitated by heat radiating off the ground after sunset. In contrast the steady air of summer will often be humid, and its very stillness means that particulate matter can just hang there messing up your images. It is a rare night indeed that is both still and clear! There are various scales used for transparency, the following is based on that used by the AAAA:

(0) The sky is overcast or raining, and no observing is possible. Entirely normal if you have just bought a new telescope.

(1) More or less completely overcast with only a few poor patches of thinner cloud or hazy open sky though which the Moon, bright planets and perhaps a few of the brightest stars (like Sirius or Capella) can be glimpsed. Not worth going outside to observe, what the AAAA calls "very poor" transparency.

(2) Incomplete cloud cover or approximately open skies but with thick haze. Visibility of Moon will be okay for naked eye or binocular views, where craters and seas can be seen easily, albeit with streaks of cloud or haze passing in front of the Moon continuously. Bright stars and a few medium-bright stars (down to about second magnitude, such as Polaris and Kochab in Ursa Minor) will be visible as well. For the AAAA this is "poor" transparency and not really worth bothering with (except in the UK where this is often as good as it gets for weeks on end).

(3) Sky essentially devoid of low-level fluffy clouds but substantial haze apparent, or else thin, wispy cirrus clouds at high altitude. "Somewhat clear" using the AAAA scale, the

planets and bright stars are easily visible and some fainter stars as well, down to about magnitude. A popular benchmark is the visibility of three or four stars in the Little Dipper asterism, besides Polaris and Kochab, Pherkad and perhaps Epsilon Ursae Minoris should be apparent as well.

(4) Fairly clear with only a little haze and fainter stars, below fourth magnitude, should be easily visible. In the Little Dipper, expect to see four or five stars, the four mentioned so far plus Zeta Ursae Minoris. Close by Kochab, 5 Ursae Minoris will be visible as well. By the AAAA scale this is termed a "partly clear" night. This, and the preceding category, is typical of the warm, humid summer nights that offer excellent seeing but poor transparency.

(5) An AAAA "clear" night, with no clouds and very little haze. With averted (indirect) vision the Milky Way should be easy enough to detect, and all the stars in the Little Dipper asterism except the faintest, Eta Ursae Minoris, can be seen.

(6) Cloudless and almost entirely haze-free skies under which the Milky Way is easily seen with direct vision. Deep sky objects like M31 and the Double Cluster in Perseus should be visible as distinct blurs to the naked eye. All seven stars in the Little Dipper can be seen. Termed a "very clear" night by the AAAA.

(7) Completely cloudless, haze-free skies that are exceptionally clean of dust and other particulate matter liable to scatter light. Considered "extremely clear" by the AAAA these are the sorts of nights under which objects like the globular clusters the Great Hercules Cluster M13 and M15 in Pegasus are obvious if blurred points of light, and galaxies like the Triangulum Galaxy M33, and M 81 can be seen with the naked eye. For keen deep sky observers, the very rare nights like these are the stuff of dreams!

Estimating Sky Darkness and Light Pollution

The final set of conditions that need to be included in an observing report is the ambient sky darkness, or as far as many people observing in suburban areas, the amount of light pollution. Dust and haze reflect natural sources of light like the Sun or Moon, even after that light source has slipped below the horizon, and even in the deep desert the night sky is never truly black. But it is near human habitation that light pollution becomes a real problem. At their worst, artificial light sources such as street lamps, billboards, houses and playing fields floodlighting produce so much unwanted light that can make the night sky not black or even dark, but a sickly orange glow. Dealing with this sort of light pollution is tricky and in general its effect can at best be diminished rather than nullified. Light pollution filters can cut out some wavelengths particular to artificial sources but absent from certain deep sky objects, primarily planetary and diffuse nebulae. Using these filters reduces overall brightness though, and so they are only worth using if you have light to spare, demanding a moderately large telescope in most cases. The least demanding of these filters in this regard are the "broadband" filters, and these can be used with even small telescopes but they have only a marginal effect. "Narrowband" filters work much better but generally need a 150-mm (6 inch) telescope or larger to be worthwhile. In addition to this limitation, none of these light pollution filters has much effect on stars or anything made up of them like galaxies or clusters. The wavelengths of light given off by stars are too similar to that of artificial light sources, and screening out the light pollution invariably dims stars as well.

John Bortle produced a "dark sky scale" for the US astronomy magazine *Sky & Telescope* that has become quite popular among amateur astronomers as a simple benchmark to use for estimating sky darkness. The full thing is rather long and involves lots of criteria for each point on the scale and is best viewed at the *Sky & Telescope* web site (the link is in Appendix 1, under Astronomy Resources and Information); what follows is a summary with some of the key things to look out for.

Class 1　Excellent dark-sky site with very faint astronomical features like the Zodiacal Light being apparent. Milky Way obvious and may even cast a shadow. Faint naked eye deep sky objects like M33 is obvious, and the naked eye should be able to spot stars down to eighth magnitude. The sky is dark right down to the horizon.

Class 2　A typical truly dark site, dark enough for M33 to be glimpsed with averted vision, and the structure of the Milky Way, such as the dark patches through Scorpio and Sagittarius, to be seen clearly. The bright globular clusters like M13 are obvious bright spots, and stars down to magnitude 7.5 should be visible.

Class 3　A good rural sky, with artificial light pollution only visible at the horizon. Bright globular clusters should be visible, and it is possible to see stars as faint as seventh magnitude.

Class 4　Rural/suburban transition with obvious regions of light pollution extending above the horizon in the direction of towns and cities. The Milky Way is obvious but not structured when overhead. Naked eye limit of star visibility is about magnitude 6.5.

Class 5　Suburban skies typical of areas even 60 km (35 miles) from big cities, with obvious light pollution along the entire horizon plus some diffuse glow to the rest of the sky. The Milky Way is weakly visible if at all along the horizon and rather pale even overhead. Stars down to about magnitude six can be seen with the naked eye.

Class 6　A bright suburban sky with the Milky Way visible only directly overhead and no structure can be made out at all. Sky glow extends a significant way above the horizon, its color depending on local conditions but enough to hide all but the brightest stars and planets. Stars down to about magnitude 5.5 can be made out.

Class 7　Suburban to urban transition with substantial light pollution evident. The sky is entirely grey rather than blue or black, with bright patches on the horizon. The Milky Way is difficult to impossible to see even directly overhead. Even bright deep sky objects like M31 can only be glimpsed, and even through a telescope they are unimpressive. The limiting magnitude for naked eye stars is around five.

Class 8　City skies; more or less uniformly grey tending towards orange, and not really dark at all. Though some of the stars making up M45, the Pleiades, can be seen, M44, the Praesepe, is invisible. Some of the stars of the fainter constellations and asterisms such as Cancer and the Little Dipper are not apparent without binoculars, indicating a limiting magnitude of around 4.5

Class 9　Inner-city skies offering nothing for deep sky observers. The sky is bright, and only the brightest stars are visible, and even then not particularly outstanding. Depending on the conditions stars as dim as magnitude four might be glimpsed.

Index